FORTSCHRITTE DER PHYSIK

HERAUSGEGEBEN IM AUFTRAGE DER PHYSIKALISCHEN GESELLSCHAFT

DER

DEUTSCHEN DEMOKRATISCHEN REPUBLIK

VON F. KASCHLUHN, A. LÖSCHE, R. RITSCHL UND R. ROMPE

HEFT 6 · 1980 · BAND 28

AKADEMIE-VERLAG · BERLIN

EVP 10,– M

31728

ISSN 0015-8208

BEZUGSMÖGLICHKEITEN

Bestellungen sind zu richten

— **in der DDR** an den Postzeitungsvertrieb, an eine Buchhandlung oder an den AKADEMIE-VERLAG, DDR - 1080 Berlin, Leipziger Straße 3—4

— **im sozialistischen Ausland** an eine Buchhandlung für fremdsprachige Literatur oder an den zuständigen Postzeitungsvertrieb

— **in der BRD und Westberlin** an eine Buchhandlung oder an die Auslieferungsstelle KUNST UND WISSEN, Erich Bieber, 7000 Stuttgart 1, Wilhelmstraße 4—6

— **in Österreich** an den Globus-Buchvertrieb, 1201 Wien, Höchstädtplatz 3

— **in den übrigen westeuropäischen Ländern** an eine Buchhandlung oder an die Auslieferungsstelle KUNST UND WISSEN, Erich Bieber GmbH, CH - 8008 Zürich/Schweiz, Dufourstraße 51

— **im übrigen Ausland** an den Internationalen Buch- und Zeitschriftenhandel; den Buchexport, Volkseigener Außenhandelsbetrieb der Deutschen Demokratischen Republik, DDR - 7010 Leipzig, Postfach 160, oder an den AKADEMIE-VERLAG, DDR - 1080 Berlin, Leipziger Straße 3—4

Zeitschrift „Fortschritte der Physik"

Herausgeber: Prof. Dr. Frank Kaschluhn, Prof. Dr. Artur Lösche, Prof. Dr. Rudolf Ritschl, Prof. Dr. Robert Rompe, im Auftrag der Physikalischen Gesellschaft der Deutschen Demokratischen Republik.
Verlag: Akademie-Verlag, DDR - 1080 Berlin, Leipziger Straße 3—4; Fernruf: 22 36 221 und 22 36 229; Telex-Nr. 114420; Bank: Staatsbank der DDR, Berlin, Konto-Nr. 6836-26-20712.
Chefredakteur: Dr. Lutz Rothkirch.
Anschrift der Redaktion: Sektion Physik der Humboldt-Universität zu Berlin, DDR - 1040 Berlin, Hessische Straße 2.
Veröffentlicht unter der Lizenznummer 1324 des Presseamtes beim Vorsitzenden des Ministerrates der Deutschen Demokratischen Republik.
Gesamtherstellung: VEB Druckhaus „Maxim Gorki", DDR - 7400 Altenburg, Carl-von-Ossietzky-Straße 30/31.
Erscheinungsweise: Die Zeitschrift „Fortschritte der Physik" erscheint monatlich. Die 12 Hefte eines Jahres bilden einen Band. Bezugspreis je Band 180,— M zuzüglich Versandspesen (Preis für die DDR: 120,— M). Preis je Heft 15,— M (Preis für die DDR: 10,— M).
Bestellnummer dieses Heftes: 1027/28/6.

AN (EDV) 57618

Fortschritte der Physik 28, 285–353 (1980)

Interaction of Photons with Electrons in Dielectric Media

MIROSLAV M. NOVAK*)

Department of Theoretical Physics, University of New South Wales, Kensington, N.S.W. 2033, Australia

Abstract

The question of the field energy-momentum tensor in a medium is not new and many workers, in the past, attempted to find the answer to it. Nevertheless, there was no general agreement about the form of such a tensor, thus resulting in a confusion involving the very fundaments of physics. Although the present work uses well established theories, the underlying philosophy is completely novel. Investigations are mainly centered around the collisionless plasma as, in that state, development of the argument is most transparent.

On the basis of a simple theoretical reasoning, it is, first of all, postulated that the momentum of the photon in a plasma is given by the Minkowski expression. This hypothesis becomes more viable as it directly leads to the accepted form for the field energy density. Furthermore, utilizing the concepts of stimulated and spontaneous emission and absorption, it is confirmed that, in the equilibrium (between radiation and plasma), the transferred momentum has the value given by the Minkowski theory. However, as will be seen, this result is valid only in the region of high frequencies. Put otherwise, when conditions are such that the laws of geometrical optics apply, the momentum of the photon is, to a good approximation, given by the Minkowski theory.

In order to arrive at a more general result, valid in the domain of wave optics, a similarity between the dispersion relation (describing the wave propagation through a medium) and the relativistic energy-momentum of a particle moving through vacuum, is explored. Accepting the equivalence of these two relations implies that some of the properties of radiation in a medium, can also be described by a massive particle travelling through empty space. In other words, the task of analysing the behaviour of electromagnetic waves in a medium, can be replaced by the analysis of the free neutral vector meson field. Adoption of this equivalence results in the field energy-momentum tensor being symmetrical.

A thorough study of the field theory then provides the basis for interpreting the Abraham tensor as the sum of the "orbital" and "spin" tensors. The former is directly connected with the energy transport, whereas the latter one is not. Realising that the magnitude of the spin term increases towards the lower end of the frequency spectrum provides the resolution of the Abraham-Minkowski dilemma: the energy-momentum tensor, corresponding to the electromagnetic wave in a medium, is that given by Abraham, while the one of Minkowski is only an approximation valid at high frequencies.

Although this conclusion stems from studies involving collisionless isotropic plasma, it is in excellent agreement with the experimental data.

---

*) Present address: Department of Natural Philosophy, University of Glasgow, Glasgow 12800, United Kingdom

Contents

I. Introduction 286
II. Covariant Electromagnetism 291
III. The Energy-Momentum Tensor in a Medium 298
1. Origin of the confusion 298
2. Theoretical considerations 302
3. Experimental results 319
IV. Electromagnetic Waves in a Plasma 322
V. Dual Nature of Radiation 332
VI. The Proca Equations 340
VII. Conclusion 349
Bibliography 351

I. Introduction

The idea that light has purely undulatory character was generally accepted in the last century. The corresponding theory was able to explain empirical facts quite well and, thus, superseded the older, corpuscular theory of Newton. As the new experimental data became available, the theory was altered, but this change was only gradual. However, towards the end of that century, the understanding of the physical phenomena was, once again, thrown into confusion. There was an accumulation of the experimental evidence which could not be accounted for by the existing theory.

Thus, Planck, in his work on the black body radiation, was forced to introduce the discontinuous aspect of radiation, in order to resolve this discrepancy between theory and experiment. This meant that the corpuscular nature of radiation was reintroduced. Einstein used and further expanded this concept, when investigating the photoelectric effect. The name "photon" was introduced to denote these corpuscules. Nevertheless, this did not mean that the corpuscular (photon) theory became (as was the case in the time of Newton) the dominant one. As distinct from the exclusiveness of the previous theories, the radiation was now to be understood as having two distinct, complementary aspects: undulatory and corpuscular. In other words, depending on the question posed, the radiation-induced effects could be accounted for by either the wave or the particle theory. The complementarity of these different characters of radiation is now well established. For instance, the classical wave theory can explain neither the photoelectric phenomenon nor the Compton effect, because these apparently belong to the particle domain. On the other hand, the occurrence of interference, diffraction, and polarization manifest the wave nature of radiation and, therefore, cannot be explained by the photon theory. Thus, in order to really comprehend the radiation phenomenon, this duality must be inevitably accepted.

In spite of differences, there is a close connection between the undulatory and corpuscular pictures. Classically, the notion of frequency and wavelength are inherent in the former, while energy and momentum belong to the latter. However, it is now customary to associate with the electromagnetic wave of angular frequency ω, a particle with energy E given by

$$E = \hbar\omega, \tag{1.1}$$

where $2\pi\hbar = h$ is the Planck constant. Although (1.1) was proposed by Planck already at the beginning of this century and is now well accepted, its meaning as to a physical connection between the energy and the frequency remains rather obscure [1]. Nevertheless, (1.1) in conjunction with one of the most fundamental principles of the special theory of relativity, viz. the principle of the equivalence of mass m and any kind of energy [2], $E = mc^2$, suggest that an effective mass $\hbar\omega/c^2$ can be attached to the photon. Now, according to the special theory of relativity, the energy E of a particle is given by

$$E^2 = p^2c^2 + m_0{}^2c^4, \tag{1.2}$$

where p stands for the momentum and m_0 refers to the rest mass of a particle. Of course, the rest mass vanishes in the case of a photon, hence its momentum has the value $\hbar\omega/c$. In addition, there is another, equivalent expression for the photon's momentum. Thus, the substitution of the relation between the frequency ν and the corresponding wavelength λ, which in vacuo takes the form

$$\nu\lambda = c, \tag{1.3}$$

together with (1.1) into (1.2) results in $p = h/\lambda$, or

$$p = \hbar k, \tag{1.4}$$

where $k = 2\pi/\lambda$ is the wave vector. Hence, (1.1) and (1.4) emphasize the relation between the corpuscular and the wave aspects of radiation. It is noticed that the ratio between the two is given in terms of the quantum of action. This, in essence, reflects the wave-corpuscular parallelism, which Einstein applied to particles with zero rest mass — photons.

The relations (1.1) and (1.4) remain, however, formally correct also for corpuscules with non-vanishing rest mass, as was first recognized by DE BROGLIE [3]. He noted the similarity between the phase of a plane wave, $\psi(\boldsymbol{r}, t) = A_0 \exp\big(i(Et - \boldsymbol{p} \cdot \boldsymbol{r})/\hbar\big)$, and the Hamiltonian action of the particle. Thus, he deduced that the Fermat principle, valid in the domain of geometrical optics, is analogous to the Maupertuis theorem in the analytical mechanics. Since the Fermat theorem has no meaning in the realm of the wave optics, he concluded that the traditional mechanics is also an approximation with the same range of validity. Therefore, de Broglie applied the wave concept of radiation to matter and showed that (1.1) and (1.4) are, indeed, relativistically invariant also for massive particles.

Subsequently, utilizing the definitions for the phase c_p and the group v_g speeds, given by the wave theory as

$$v_p = \omega/k \qquad v_g = d\omega/dk, \tag{1.5}$$

he arrived at the well known group speed theorem [4]. In view of its derivation, it is valid in the geometrical optics approximation and expresses the relation between the speed v of a particle and the phase speed v_p of the associated wave — the so called matter wave

$$v_p = c^2/v. \tag{1.6}$$

It is then easily shown [4] that the group speed of the ψ (matter) waves connected with the free particle is equal to the speed of the latter. Hence, a photon travelling in vacuum appears as a particle whose speed (and also group speed) is equal to the phase speed of the corresponding wave, as expected.

Despite these similarities, massless photons and massive particles are not the same. Thus, in order for a particle to be observed, it must interact. Consequently, its energy and momentum will be changed in accordance with the conservation laws. Then, high energy material particles are like high energy photons, because they change energy by changing their effective mass. This is the domain of the applicability of the geometrical optics, which, by definition [5], is valid in the region of small wavelengths. On the other hand, the differences between photons and particles are very pronounced at the other end of the energy spectrum, i.e., in the low energy limit. The energy of material particles approaches, in this case, the rest energy, whereas the photon energy (and frequency) approaches zero. Therefore, in this limiting case the theory of radiation is effectively replaced by the static field theory. Thus, classically, our conception of matter is represented by the particle nature, while that of radiation assumes the wave character.

It is important to remember this differentiation related to the energy spectrum, as it will, in the following, provide one of the crucial points of the analysis.

The main investigation of the present work is directed towards the electromagnetic energy-momentum tensor in a medium. This tensor is very important as it provides the basic information related to a system in question. It is directly involved in the various conservation laws. When written out explicitly, its components can be identified as follows: the time-time component is equal to the field energy density, the space-time components describe the Poynting vector and the momentum density, and its space-space components comprise the electromagnetic stress tensor.

The form of the energy-momentum tensor corresponding to a field in vacuo is well established. However, when such a field is embedded in a medium, the form of the tensor is, generally, not agreed upon. In view of its great significance, this may come as a surprise to many, especially when it is realized that the principal work on the subject was done more than sixty years ago. The distinction between the various proposed tensors is rather subtle and the resolution of the existing confusion is by no means trivial. As there is scarcely any coverage of this dilemma in the books, the large majority is not even aware of the existence of the problem.

Amongst those who studied the topic, opinions differ. Their research is exclusively directed towards the dielectric medium with the refractive index n larger than unity. Complexity of the problem is reflected in many attempts by various workers to find the energy-momentum tensor of a field in a medium. Thus, the first proposition put forward was due to Minkowski. Making use of the macroscopic Maxwell equations in the energy and momentum conservation laws, he arrived at a field energy-momentum tensor which was not symmetrical. This lack of symmetry was rather disturbing as it led to some peculiar consequences. Therefore, Abraham (and Hertz) suggested a symmetrized version. As a consequence, the value of the momentum density differed now by the factor of n^2. Both derivations seems to be based on sound physical principles, yet they lead to different conclusions. It is important to find the correct expression as it is closely connected with the concept of the ponderomotive force. Although both yield the same magnitude for this force, the directions predicted by them are mutually opposing. Thus, a number of workers set out to resolve this controversy. In his original article on relativity, PAULI [*2*] appeared to favour the Abraham tensor but, later, apparently being convinced by the argument due to VON LAUE [*6*], reverted to the Minkowski proposition. That argument required that the velocities of the energy propagation (in a medium) must satisfy the Einstein addition of velocity theorem, and only Minkowski tensor was able to fulfil this condition. Hence, the von Laue criterion had a profound influence on further developments regarding the field energy-momentum tensor. This question was also analyzed by MOLLER [7] and he opted too (apparently) for the non-symmetric (Minkowski) tensor. On the other hand, LANDAU and LIFSHITZ [*8*] adopt the version due to Abraham. In fact, most authors now tend to accept for the electromagnetic energy-momentum tensor in a dielectric the form given by Abraham. Amongst these are BALAZS [*9*], SKOBEL'TSYN [*10*], BURT and PEIERLS [*11*], GORDON [*12*], PEIERLS [*13*], JONES [*14*], and SHOCKLEY [*15*]. Some others, like e.g., HAUS [*16*] and ROBINSON [*17*], advocate the necessity of additive correction terms to the Abraham tensor. Furthermore, BREVIK [*18*], GINZBURG [*19*], and MIKURA [*20*] advance arguments intended to show the equivalence of both proposed tensors.

To complicate matters even further, other forms for the electromagnetic energy-momentum tensor have been put forward as well. Thus, EINSTEIN and LAUB [21] examined the problem and arrived at a completely different result from those of MINKOWSKI and ABRAHAM. Another proposal is due to BECK [22], who arrived at a conclusion similar to that of MARX and GYÖRGYI [23].

From the in depth analysis of the properties of a field in a medium, DE GROOT and SUTTORP [24] obtained two different expressions for the energy-momentum tensor. The first one was derived by the statistical averaging of the relevant microscopic quantities, while the second one was found by subtracting the total energy-momentum in the absence of the field from the one when the field was present.

It would appear that this confused situation should be easy to resolve experimentally. This, however, is not so. Results obtained from the few experiments actually performed do not correlate. Thus, the data collected by JONES and RICHARDS [25], and in the more accurate, recent version of the same experiment by JONES and LESLIE [26], lead to the conclusion in agreement with the Minkowski form of the momentum. This finding is also supported by the experiment of ASHKIN and DZIEDZIC [27]. Nevertheless, the empirical data from WALKER et al. [28] and WALKER and WALKER [29] suggest that the photon momentum in a medium is given by the Abraham value. In addition, results obtained in the static limit by HAKIM and HIGHAM [30] confirm the Abraham theory.

On the basis of the brief description of the theoretical and experimental results, it is seen that the confusion is rather profound. The obvious difficulty is partly due to the complexity of the question of the nature of the photon momentum in a medium. Furthermore, there is a rather large variation in the conditions under which experiments were performed. For instance, in the Jones and Richards experiment, visible light ($\nu \approx 10^{15}$ Hz) was used and liquid under investigation had the refractive index of the order of unity. On the other hand, Walker et al. employed very low frequency ($\nu < 100$ Hz) and a ceramic sample had very high dielectric constant (≈ 3200). It is, therefore, not surprising that such a wide range of the involved variables produces contradictory results.

In order to comprehend this, rather peculiar, behaviour, it is necessary, first of all, to consider the simplest case. As is well known, a medium can be in any of the four states: solid, liquid, gas, and plasma. Each of these states is characterized by the different set of variables. Therefore, in order to simplify the structure and to emphasize the physical nature of this fundamental problem, we shall concentrate our attention on the case of the plasma state. It will be assumed that there is no external magnetic field present and that collisions among electrons are rare and, hence, can be neglected. In other words, our main effort will be directed toward an isotropic, collisionless plasma with the dispersion relation of the form

$$\varepsilon(\omega) = 1 - (\omega_p^2/\omega^2), \tag{1.7}$$

where $\varepsilon(\omega)$ is the frequency dependent dielectric permeability, $\omega_p^2 = 4\pi NZq^2/m$ is the plasma frequency, ω denotes the frequency of the incident wave, and NZ, q, m represent the total number density, charge, and mass of the electrons, respectively. The reason for this selection is due to the weak nature of the interaction among plasma particles. This, consequently, considerably simplifies the procedure. Once this relatively simple case has been well understood, the extension to other states is then more likely to lead to the correct conclusion. Furthermore, most of the results obtained from the investigation of a tenuous plasma will also apply for the case of dielectrics and metals when the incident frequency is approximately in (or above) the ultraviolet range, since (1.7) may be shown [8] to be valid at high frequencies for matter in any state.

As may be deduced by now, the central task of the present work involves the analysis of the interaction of an electromagnetic field with a dielectric medium. Hence, to facilitate the inquiry, the following scheme is proposed: The field theory, which provides the necessary background information for this type of problem, is briefly outlined in the next chapter. The four-dimensional formalism of the special theory of relativity is used throughout that chapter. Starting from the covariant form of the Maxwell equations, the energy momentum tensor of the total system is derived employing the concept of the Lagrangian density. It is then found that part of this tensor refers to the electromagnetic field, while the other part describes the background medium. This "splitting-up" procedure of the total tensor is, however, not unique. As a consequence, this is responsible for the above ambiguity in selecting the form of the field energy-momentum tensor.

A comprehensive review and an analysis of the various proposed forms of that tensor are presented in Chapter III. As already indicated, they lead to different conclusions which are then also examined. In addition, the correlation between the theory and existing empirical facts is discussed.

As mentioned above, the present analysis is centered on the fourth state of matter, i.e., plasma. There are various kinds of electromagnetic waves that such a state can support. It is, therefore, instructive to include some details about the wave propagation and also to recall the essential features of the supporting medium. The problem of defining dielectric permittivity of such a system is a rather complex one and it has attracted much attention [*31, 32, 33*]. These questions are treated in Chapter IV.

Elementary considerations, which often help in understanding the dual nature of radiation are examined in Chapter V. It will be found that, within the WKB approximation, this simple theory is in agreement with the accepted results. The de Broglie group velocity theorem is utilized in order to arrive at some conclusions of that chapter. Further, it is of interest to inquire into the momentum transfer from the radiation field to a particle moving in a plasma. The method used by EINSTEIN [*34*] in analysis of the black body radiation is, first of all, adapted to include the media with the refractive index differing from unity, along the lines attempted by SKOBEL'TSYN [*35*]. It is then applied in derivation of the momentum transfer, thus directly leading to the accepted value [*36, 37, 38*] of the radiation energy density inside plasma. Let us emphasize again that these results are applicable to the case of an isotropic, collisionless plasma. A possible extension to the collisional plasma is also suggested, in analogy to the treatment of a nonlinear force by HORA [*39*].

It was pointed out by COLE [*40, 41*], LEE [*42*], and later by KLIMA and PETRZILKA [*43*] that the dispersion equation for the transverse electromagnetic waves in plasma is formally identical with the energy-momentum relation of a massive particle. In other words, this suggests that the photon, with vanishing rest mass in vacuo, when propagating in a plasma, acquires a non-zero effective rest mass. Put otherwise, a photon in a plasma behaves as a neutral vector meson in vacuo. Hence, in order to resolve the energy-momentum controversy, it is instructive to resort to the Proca field. The electrodynamics for such a field is developed by modifying the results of classical electrodynamics. It leads to a tensor which is symmetrical [*44, 45*]. A careful analysis of the work by BELINFANTE [*46, 47*] shows that the symmetrizing term, which is normally added to the canonical energy-momentum tensor, may be interpreted as the spin angular momentum. As the time-time and space-time components of the canonical and symmetric energy-momentum tensors are the same [*48*], it is deduced that the spin angular momentum is responsible for the additional linear momentum which, however, is not connected with the energy transport — in accordance with a similar proposal put forward by BOPP [*49*].

The concluding chapter of the present work briefly reviews the main findings of this investigation. It is argued there that the complete field energy-momentum tensor is, in fact, that proposed by Abraham with the Minkowski version being only approximately valid at high frequencies.

II. Covariant Electromagnetism

The notion of the energy-momentum tensor is closely linked with the field equations. Thus, it is appropriate to recall these equations at this stage and to show how such a tensor is derived. The special theory of relativity is used throughout, as its formalism is most convenient. By utilizing only four-scalars, four-vectors, and tensors, the theory clearly manifests the Lorentz invariance. In other words, it is form independent under the Lorentz transformations.

The summation convention, i.e., summation over the repeated indices will be employed. Furthermore, Greek indices run from 0 to 3, while Roman ones from 1 to 3. In accord with this convention, the time coordinate of the position four-vector becomes $x^0 = ct$, while space components are denoted by (x_1, x_2, x_3). A four-vector position can, thus, be expressed concisely as $x^\alpha = (ct, \boldsymbol{x})$. The metric tensor is diagonal and its components are $g^{00} = 1$, $g^{11} = g^{22} = g^{33} = -1$. The quantities denoted by a prime will refer to the inertial frame which moves with a uniform velocity with respect to the unprimed frame. Hence, the four-vectors, as measured in both frames, are connected by a 4×4 matrix, $a^{\mu\nu}$, as

$$A^{\mu'} = a^{\mu\nu} A^\nu, \tag{2.1}$$

i.e., the elements of a four-vector transform like the coordinates. Similarly, a tensor of the rank two transforms as

$$A^{\mu\nu'} = a^{\mu\lambda} a^{\nu\eta} A^{\lambda\eta}. \tag{2.2}$$

We are now in a position to consider the electromagnetic field. As is well known [*50*], the scalar potential φ and the vector potential $\boldsymbol{A}$ are the components of a four-potential A^α. Although they are not directly measurable, and are introduced in order to simplify the calculations, their direct relation to the observed field strengths is well established

$$\boldsymbol{E} = -\frac{1}{c}\,\partial_t \boldsymbol{A} - \nabla\varphi \qquad \boldsymbol{H} = \nabla \times \boldsymbol{A}. \tag{2.3}$$

When written in the component form, they become

$$\begin{aligned} E_x &= -(\partial^0 A^1 - \partial^1 A^0) \qquad & H_x &= -(\partial^2 A^3 - \partial^3 A^2) \\ E_y &= -(\partial^0 A^2 - \partial^2 A^0) \qquad & H_y &= -(\partial^3 A^1 - \partial^1 A^3) \\ E_z &= -(\partial^0 A^3 - \partial^3 A^0) \qquad & H_z &= -(\partial^1 A^2 - \partial^2 A^1), \end{aligned}$$

where $\partial_\alpha \equiv \partial/\partial x^\alpha$.

It is convenient to combine these components in a matrix form. Thus, defining the antisymmetric field tensor $F^{\mu\nu}$ as

$$F^{\mu\nu} = \partial^\mu A^\nu - \partial^\nu A^\mu, \tag{2.4}$$

we obtain the matrix expressed in terms of the field strengths

$$F^{\mu\nu} = \begin{pmatrix} 0 & -E_x & -E_y & -E_z \\ E_x & 0 & -H_z & H_y \\ E_y & H_z & 0 & -H_x \\ E_z & -H_y & H_x & 0 \end{pmatrix}. \tag{2.5}$$

In terms of the field tensor $F^{\mu\nu}$, the homogeneous Maxwell equations take the form

$$\partial^\lambda F^{\mu\nu} + \partial^\mu F^{\nu\lambda} + \partial^\nu F^{\lambda\mu} = 0. \tag{2.6}$$

These equations can also be expressed through the dual field tensor $\mathcal{F}^{\mu\nu}$, which is defined as

$$\mathcal{F}^{\mu\nu} = \frac{1}{2} \varepsilon^{\mu\nu\varkappa\sigma} F_{\varkappa\sigma}, \tag{2.7}$$

where $\varepsilon^{\mu\nu\varkappa\sigma}$, the antisymmetric tensor, is equal to

1 for any even permutation of indices,

−1 for any odd permutation,

0 if two indices are equal.

The homogeneous field equations in terms of the dual field tensor assume the form

$$\partial_\mu \mathcal{F}^{\mu\nu} = 0. \tag{2.8}$$

Mathematically speaking, (2.6) and (2.8) are equivalent. However, as in an ordinary space the electric field vector $\boldsymbol{E}$ is polar and that due to the magnetic field $\boldsymbol{H}$ is axial, it seems more correct [*2*] to represent the electromagnetic field by $F^{\mu\nu}$.
The inhomogeneous pair of the Maxwell equations can also be written in the relativistically invariant form. Thus, in terms of $F^{\mu\nu}$ and the four-current j^μ, these take on the form

$$\partial_\mu F^{\mu\nu} = \frac{4\pi}{c} j^\nu. \tag{2.9}$$

The space components of the four-current, $\boldsymbol{j} = \varrho \boldsymbol{v}$, form a vector in three dimensions, the so called current density vector. The time component is the product of the charge density ϱ and the vacuum speed of light c. The relation between these components of the four-current is expressed through the continuity equation

$$\partial_\mu j^\mu = 0. \tag{2.10}$$

In a continuous medium we think of the charges as smeared over space-time manifold. Four-current then takes this effect into account. Of course, if there are no charges present, the right hand side of (2.9) vanishes. In the macroscopic formulation it is, however, necessary, because of polarization and magnetization effects, to introduce the additional field tensor $H^{\mu\nu}$. This tensor is obtained from $F^{\mu\nu}$ by the transformation $\boldsymbol{E} \to \boldsymbol{D}$ and $\boldsymbol{H} \to \boldsymbol{B}$. The covariant form of the macroscopic Maxwell equations is then

$$\partial_\mu \mathcal{F}^{\mu\nu} = 0 \qquad \partial_\mu H^{\mu\nu} = \frac{4\pi}{c} j^\nu. \tag{2.11}$$

The definitions of the four-current j^μ, the four-potential A^μ, and the field tensors $F^{\mu\nu}$ and $H^{\mu\nu}$, in conjunction with the Maxwell equations conclusively establish the covariance of the field equations [*50*].
Turn attention now to the covariant formulation of the conservation laws. It will be assumed here that the system under investigation consists of an electromagnetic field and that there also are some charged particles moving in it. Hence, we shall examine the interaction between these two subsystems, resorting to the Lagrangian formalism. Of prime importance, in the description, is the action integral A, defined as [*5*]

$$A = \int \mathcal{L} \, d\tau, \tag{2.12}$$

where the four-dimensional volume element is

$$d\tau = c \, dt \, dx \, dy \, dz \tag{2.13}$$

and $\mathscr{L}$ refers to the Lagrangian density. This function, as applied to our system, consists of three contributions

$$A = A_f + A_m + A_{fm}. \tag{2.14}$$

The first term, A_f, refers to the action of the field alone, i.e., to system in the absence of any charges. If one knows this term, the field equations can be easily [5] obtained. The Lagrangian density corresponding to this term must be quadratic in the field, in order that the field equations remain linear. Furthermore, as this function must be Lorentz invariant, it cannot involve coordinates explicitly. In addition, it cannot be expressed in terms of the four-potentials, as these are not uniquely determined. In view of these restrictions, the only Lorentz invariant quadratic forms are $F_{\mu\nu}F^{\mu\nu}$ and $F_{\mu\nu}\mathscr{F}^{\mu\nu}$. However, the latter scalar changes sign under the inversion. Therefore, only the former scalar, $F^{\mu\nu}F_{\mu\nu}$, constitutes the free field Lagrangian. Thence, the action A_f, corresponding to the free field, takes the form

$$A_f = -\frac{1}{16\pi c}\int F_{\mu\nu}F^{\mu\nu}\,d\tau\,, \tag{2.15}$$

which, in the three dimensional representation, becomes

$$A_f = \frac{1}{8\pi}\int (E^2 - H^2)\,dV\,dt. \tag{2.15a}$$

In order to find the Lagrangian density for the free field, it is only necessary to compare (2.12) with (2.15a). One then obtains

$$\mathscr{L}_f = \frac{1}{8\pi}(E^2 - H^2). \tag{2.16}$$

The second term in (2.14), A_m, represents that part of the action which depends exclusively on the properties of the particles, i.e., the system in the absence of the field. If there are several particles present, then the general form for the action is [5]

$$A_m = -\sum mc \int ds. \tag{2.17}$$

In other words, A_m is equal to the sum of the actions referring separately to each particle. The last term in (2.14), A_{fm}, describes the interaction between the components of our system, viz. between the field and matter. Therefore, it must contain information regarding both subsystems. Hence, it is found convenient to represent matter by the electronic charge, q, while the field is determined by the four-potential, A^μ. The action term then takes the form

$$A_{fm} = -\sum \frac{q}{c}\int A_\mu\,dx^\mu. \tag{2.18}$$

If discrete charges are replaced by the continuous charge density ϱ, then (2.18) assumes the form

$$A_{fm} = -\frac{1}{c^2}\int A_\mu j^\mu\,d\tau. \tag{2.19}$$

Combining (2.15), (2.17), and (2.19), the action for the whole system (with a continuous charge distribution) then becomes

$$A = -\frac{1}{16\pi c}\int F_{\mu\nu}F^{\mu\nu}\,d\tau - \sum\int mc\,ds - \frac{1}{c^2}\int A_\mu j^\mu\,d\tau. \tag{2.20}$$

With this in mind, the electromagnetic Lagrangian density becomes

$$\mathcal{L} = -\frac{1}{16\pi} F_{\mu\nu} F^{\mu\nu} - \frac{1}{c} j_\mu A^\mu . \tag{2.21}$$

Having found the Lagrangian density, the field equations can be readily obtained. Thus, utilizing the Euler-Lagrange equations of motion

$$\partial^\nu \frac{\partial \mathcal{L}}{\partial(\partial^\nu A_\mu)} - \frac{\partial \mathcal{L}}{\partial A_\mu} = 0, \tag{2.22}$$

together with (2.21), one arrives at the field equations. To elucidate the meaning of (2.22), the required operations are first performed. After some calculation, we obtain

$$\frac{1}{4\pi} \partial^\nu F_{\mu\nu} + \frac{1}{c} j_\mu = 0. \tag{2.23}$$

As is easily recognized, these are nothing else but the inhomogeneous Maxwell equations (2.9). To obtain the conservation of the current density, we proceed to take the divergence of (2.23), thus yielding

$$\frac{1}{4\pi} \partial^\mu \partial^\nu F_{\mu\nu} + \frac{1}{c} \partial^\mu j_\mu = 0.$$

However, it is recalled that the product of a symmetric operator $\partial^\mu \partial^\nu$ and the anti-symmetric tensor $F_{\mu\nu}$ vanishes, resulting in

$$\partial^\mu j_\mu = 0, \tag{2.24}$$

the earlier obtained continuity equation.

We are now in a position to present the derivation of an expression for the energy and momentum of the electromagnetic field. In doing so, it is found convenient to utilize the Hamiltonian formalism. Thus, in the particle mechanics the canonical momentum p_i is defined by

$$p_i = \frac{\partial \mathcal{L}}{\partial \dot{q}_i} \tag{2.25}$$

and the Hamiltonian density $\mathcal{H}$ then becomes

$$\mathcal{H} = \frac{\partial \mathcal{L}}{\partial \dot{q}_i} \dot{q}_i - \mathcal{L}, \tag{2.26}$$

where $\dot{q}_i$ refers to the time derivative of the generalized coordinates q_i. Similarly, the field canonical momentum is obtained from (2.25) by the appropriate replacement for $\dot{q}_i$. Thus, it is then defined by

$$p_i = \frac{\partial \mathcal{L}}{\partial(\partial^\mu A_\nu)}, \tag{2.27}$$

what means that the field momentum is canonically conjugate to A_ν. In analogy, the expression for the field Hamiltonian density can be obtained. The result is then known as the canonical energy-momentum tensor [*48*]

$$T^{\mu\nu} = \frac{\partial \mathcal{L}}{\partial(\partial_\mu A^\varkappa)} \partial^\nu A^\varkappa - g^{\mu\nu} \mathcal{L}. \tag{2.28}$$

In general, this tensor is not symmetric. Furthermore, the component corresponding to the energy density is not always positive [*48*]. These facts have caused some difficulties. To make this tensor more transparent, consider first the free electromagnetic field. The second term on the right hand side of (2.21) then vanishes (as there are no charges present). The free field Lagrangian, $\mathscr{L} = -1/16\pi\, F_{\mu\nu}F^{\mu\nu}$, is then substituted in (2.28) to give

$$T^{\mu\nu} = -\frac{1}{4\pi}\, g^{\mu\alpha}F_{\alpha\,\lambda}\partial^{\nu}A^{\lambda} + \frac{1}{16\pi}\, g^{\mu\nu}F_{\alpha\beta}F^{\alpha\beta}. \tag{2.29}$$

This expression becomes clearer when it is written in the three-dimensional form. Thus, substituting (2.5), the matrix form for the field tensor $F^{\mu\nu}$, together with the free field Lagrangian (2.16) into (2.24) gives us the following components for the electromagnetic energy-momentum tensor

$$\begin{aligned} T^{00} &= \frac{1}{8\pi}\,(E^2 + H^2) + \frac{1}{4\pi}\, \nabla \cdot (\varphi \boldsymbol{E}) \\ T^{0i} &= \frac{1}{4\pi}\,(\boldsymbol{E}\times\boldsymbol{H})_i + \frac{1}{4\pi}\, \nabla \cdot (A_i E) \\ T^{i0} &= \frac{1}{4\pi}\,(\boldsymbol{E}\times\boldsymbol{H})_i + \frac{1}{4\pi}\,[(\nabla \times \varphi \boldsymbol{H})_i - \partial_0(\varphi E_i)]. \end{aligned} \tag{2.30}$$

It is seen that all expressions differ from the commonly accepted ones for the energy and momentum densities by the addition of the last term. This difficulty, however, is usually obviated. It is recalled that the components of the energy-momentum tensor refer to the densities. However, in order to obtain energy and momentum of the field, they must be integrated over all space. As it is assumed that the fields are localized, the contributions from the added terms in (2.30) vanish when the integration is performed, thus leading to the standard expression for the field energy and momentum.

Knowledge of the energy-momentum tensor enables one to write down the conservation laws. Thus, the Poynting theorem, which expresses the conservation of energy and momentum, can be written [*2*] in the covariant form as

$$\partial_\mu T^{\mu\nu} = 0. \tag{2.31}$$

In analogy, the vanishing of the four-divergence of the angular momentum density implies that the orbital angular momentum of the field is conserved. That is, if the angular momentum tensor is

$$M_T{}^{\mu\nu\varkappa} = T^{\mu\nu}x^{\varkappa} - T^{\mu\varkappa}x^{\nu}, \tag{2.32}$$

then

$$\partial_\mu M_T{}^{\mu\nu\varkappa} = 0 \tag{2.33}$$

expresses the appropriate conservation law. However, in order for (2.33) to hold, the direct calculation requires that $T^{\mu\nu}$ must be symmetric. As was already noticed, and is also clearly exhibited by comparing T^{0i} and T^{i0} in (2.30), the tensor $T^{\mu\nu}$ does not satisfy this symmetry requirement. Hence, the angular momentum, with $M_T{}^{\mu\nu\varkappa}$ as defined by (2.32), is not conserved. In addition, according to the definition of $T^{\mu\nu}$, (2.29), it depends explicitly on the potentials. However, it is known that the potentials are not uniquely determined. Consequently, the tensor $T^{\mu\nu}$ is not gauge invariant [*50*]. Moreover, direct calculation shows that its trace, $T^{\mu\nu}$, does not vanish.

As the just considered field is free (in view of (2.16)), it is difficult to accept the tensor $T^{\mu\nu}$ as physically significant. Therefore, it was necessary to find another energy-momentum tensor, which would remove the above mentioned deficiencies. Here lies the origin of the symmetric energy-momentum tensor $\Theta^{\mu\nu}$. The prescription for such a tensor was first given by BELINFANTE [46, 47]. Thus, to symmetrize $T^{\mu\nu}$, the term $\partial^\nu A^\lambda$ in (2.29) is replaced, utilizing (2.4), by $\partial^\lambda A^\nu - F^{\lambda\nu}$. Then, the tensor $T^{\mu\nu}$ assumes the form

$$T^{\mu\nu} = \frac{1}{4\pi}\left[g^{\mu\alpha}F_{\alpha\lambda}F^{\lambda\nu} + \frac{1}{4}\,g^{\mu\nu}F_{\alpha\beta}F^{\alpha\beta}\right] - \frac{1}{4\pi}\,g^{\mu\alpha}F_{\alpha\lambda}\,\partial^\lambda A^\nu. \tag{2.34}$$

The last term in above can now be rewritten as

$$T_s^{\mu\nu} = -\frac{1}{4\pi}\,g^{\mu\alpha}F_{\alpha\lambda}\,\partial^\lambda A^\nu = \frac{1}{4\pi}\,F^{\lambda\mu}\,\partial_\lambda A^\nu. \tag{2.35}$$

It is assumed that the field is free. That is, as there are no charges present, the Maxwell equations have the form (2.9), $\partial_\lambda F^{\lambda\mu} = 0$. Thus, a term $A^\nu\partial_\lambda F^{\lambda\mu}$ can be added to (2.35). Consequently, we can write

$$T_s^{\mu\nu} = \frac{1}{4\pi}\,\partial_\lambda(F^{\lambda\mu}A^\nu). \tag{2.36}$$

The symmetric energy-momentum tensor $\Theta^{\mu\nu}$ is now defined as

$$\Theta^{\mu\nu} = T^{\mu\nu} - T_s^{\mu\nu}$$

or, written in terms of the field tensors it becomes

$$\Theta^{\mu\nu} = \frac{1}{4\pi}\left[g^{\mu\alpha}F_{\alpha\lambda}F^{\lambda\nu} + \frac{1}{4}\,g^{\mu\nu}F_{\alpha\beta}F^{\alpha\beta}\right]. \tag{2.37}$$

It is clear that the potentials are now not explicitly involved in the definition, therefore, the tensor $\Theta^{\mu\nu}$ is gauge invariant [5]. If the field energy tensor (2.5) is substituted in (2.37), then $\Theta^{\mu\nu}$ can be written in the component form as

$$\begin{aligned}\Theta^{00} &= \frac{1}{8\pi}\,(E^2 + H^2)\\ \Theta^{0i} &= \frac{1}{4\pi}\,(\boldsymbol{E}\times\boldsymbol{H})_i\\ \Theta^{ij} &= -\frac{1}{4\pi}\left[E_iE_j + H_iH_j - \frac{1}{2}\,\delta_{ij}(E^2 + H^2)\right].\end{aligned} \tag{2.38}$$

It is immediately obvious that the expression for the energy and momentum densities given by the time-time and space-time components of $\Theta^{\mu\nu}$, respectively, are in accordance with the accepted formulae. Furthermore, the space-space components represent the Maxwell stress tensor taken with the negative sign.

The four-divergence of $T_s^{\mu\nu}$ vanishes as it is the product of symmetrical and antisymmetrical tensors. Therefore, in view of its derivation, the tensor $\Theta^{\mu\nu}$ also has vanishing divergence. Expressed mathematically,

$$\partial_\mu\Theta^{\mu\nu} = 0. \tag{2.39}$$

Of course, (2.39) is nothing else but the covariant form of the conservation of energy and momentum. In other words, it is the four-dimensional equivalent of the Poynting theorem. It is to be noted that the comparison of (2.29) with (2.37) yields the equality of the total energy and momentum integrals over space-like volumes [48]. Pauli has, furthermore, emphasized that for the unique localization in space-time of the energy and momentum, the gravitational theory is necessary. The introduction of the gravitational field gives, then, an unambiguous meaning to the energy-momentum tensor.

The total angular momentum density of the field is now defined, using $\Theta^{\mu\nu}$, in terms of the tensor $M^{\mu\nu\varkappa}$, as

$$M^{\mu\nu\varkappa} = \Theta^{\mu\nu} x^{\varkappa} - \Theta^{\mu\varkappa} x^{\nu}. \tag{2.40}$$

In view of the symmetry of $\Theta^{\mu\nu}$, the total angular momentum is conserved, which is mathematically expressed by

$$\partial_\mu M^{\mu\nu\varkappa} = 0. \tag{2.41}$$

The total angular momentum is again obtained by the integration of the corresponding density over space-like volumes.

So far, the considerations were mainly directed towards a free, i.e., non-interacting electromagnetic field. If the charges are included, interaction occurs. Consequently, the Lagrangian for the Maxwell field is also changed and is now given by (2.21). This, in turn, implies that the energy-momentum tensor $\Theta^{\mu\nu}$ is no longer divergenceless. Then, instead of (2.39), we now write

$$\partial_\mu \Theta^{\mu\nu} = -\frac{1}{c} F^{\nu\lambda} j_\lambda. \tag{2.42}$$

The right hand side of (2.42) describes the interaction between the electromagnetic field, here represented by $F^{\nu\lambda}$, and the four-current j_λ, describing the source. This coupling term is known as the Lorentz force density. The integral of this force density over the space volume then gives the time rate of change of the total four-momentum of all particles. On the other hand, the space integral of $\partial_\mu \Theta^{\mu\nu}$ describes the time rate of change of the field four-momentum. Therefore, the conservation of the momentum of the whole system (field and particles) is written as

$$\int (\partial_\mu \Theta^{\mu\nu} + f^\nu)\, dV = 0, \tag{2.43}$$

where the Lorentz force $f^\nu = 1/c\, F^{\nu\lambda} j_\lambda$. Thus, the sum of the total momentum and energy of the system is conserved.

The Poynting theorem can also be interpreted as follows: the time-time component of the Lorentz force density multiplied by c is $\boldsymbol{j} \cdot \boldsymbol{E}$. The space volume integration then describes the conversion of the electromagnetic energy into the mechanical or heat energy. Hence, it must be balanced by the corresponding decrease of the field energy. Physically, this is equal to work done by the field per unit time. Thus, it represents the rate of increase of the mechanical energy. In analogy, the considerations of a similar nature apply to the linear momentum.

Landau and Lifshitz [5] proceed along slightly different lines. First of all, they determine the field energy-momentum tensor. Then, the corresponding tensor for the particles is derived. The total energy-momentum tensor, which is equal to the sum of the energies of the field and particle tensors, then satisfies what is essentially our (2.42), as it can be written as

$$\partial_\mu (\Theta_f^{\mu\nu} + \Theta_p^{\mu\nu}) = 0. \tag{2.44}$$

The total tensor must be symmetric. From the fact that the tensor representing the particles, $\Theta_p^{\mu\nu}$, is symmetric, they conclude that also the field tensor $\Theta_f^{\mu\nu}$ must be symmetric.

The above discussion has centered about the system with the Lagrangian density having one, (2.16), or two, (2.21), terms. A more accurate treatment is presented by Rohrlich

[*45*] where the three-term Lagrangian is used, which makes it possible to investigate such effects as radiation reaction, self-energy, etc.
The derivation of the field energy-momentum tensor inside a macroscopic medium presents a complex problem. We saw above that it was possible to divide the tensor corresponding to the total system, consisting of field and particles, into two parts, each representing the separate component of the system. However, when the macroscopic medium is involved, this separation cannot be established uniquely. The main reason for the implied ambiguity lies in the impossibility to differentiate as to what is electromagnetic and what is not. The following chapter examines the different proposals put forward to avoid this dilemma.

III. The Energy-Momentum Tensor in a Medium

1. Origin of the confusion

The considerations of the preceding chapter apply, essentially, to the electromagnetic field in vacuo interacting with the charged particles. Results arising from such analysis are well established. These do not hold, however, when the field energy-momentum tensor inside a macroscopic body is discussed. Judging on the basis of a number of proposed forms for such a tensor, it is clear that neither is the question trivial nor is the involved subtlety sufficiently appreciated. It may appear surprising that a problem as fundamental as this has still not been satisfactorily resolved, although the main work on the subject was completed decades ago. Numerous articles appearing in the literature testify to the complexity of the problem. The question of the electromagnetic energy-momentum tensor was almost completely solved by Helmholtz, (cf. [*51*]), decades before the controversy originated. Using the principle of virtual work in conjunction with the macroscopic field equations, Helmholtz found the force which acts on a medium in the static field.
Amongst the early workers investigating this subject was Minkowski. By a straightforward application of the macroscopic Maxwell equations to the energy and momentum conservation laws, he arrived [*2*] at the energy-momentum tensor of the form

$$S_M{}^{\mu\nu} = \frac{1}{4\pi}\left[g^{\mu\alpha}F_{\alpha\lambda}H^{\lambda\nu} + \frac{1}{4}\,g^{\mu\nu}F_{\alpha\beta}H^{\alpha\beta}\right]. \tag{3.1}$$

Substitution of the components of the field strength tensors $F_{\mu\nu}$ and $H^{\mu\nu}$ into (3.1) results in the space-space part of this tensor being expressed in terms of the field variables, thus becoming

$$S_M{}^{ik} = \frac{1}{4\pi}\left[E_iD_k + H_iB_k - \frac{1}{2}\,(\boldsymbol{E}\cdot\boldsymbol{D} + \boldsymbol{H}\cdot\boldsymbol{B})\,\delta^{ik}\right]. \tag{3.2}$$

Recall that $F_{\mu\nu} = g_{\alpha\gamma}F^{\gamma\delta}g_{\delta\nu}$ and that $H^{\mu\nu}$ is obtained from $F^{\mu\nu}$ by substituting $\boldsymbol{E}\to\boldsymbol{D}$ and $\boldsymbol{H}\to\boldsymbol{B}$. The influence of the medium is taken into account by introducing concepts of the tensors ε_{ik} and μ_{ik} of dielectric permittivity and magnetic permeability, respectively. Hence, we write $D_i = \varepsilon_{ik}E_k$ and $B_i = \mu_{ik}H_k$.
It is generally accepted [*2*] that the time-time, $S_M{}^{00}$, component of the tensor (3.2) refers to the field energy density

$$S_M{}^{00} = \frac{1}{8\pi}\,(\boldsymbol{E}\cdot\boldsymbol{D} + \boldsymbol{H}\cdot\boldsymbol{B}), \tag{3.3}$$

while the time-space, S^{0i}, component represents the energy flux

$$S_M{}^{0i} = \frac{1}{4\pi}(\boldsymbol{E}\times\boldsymbol{H})_i = \frac{S_i}{c}, \tag{3.4}$$

where $\boldsymbol{S} = c/4\pi\,(\boldsymbol{E}\times\boldsymbol{H})$ is the well known Poynting vector.
Now, on calculating $S_M{}^{0i}$, the space-time component of the (3.1) tensor, we have

$$S_M{}^{0i} = \frac{1}{4\pi}(\boldsymbol{D}\times\boldsymbol{B})_i = \frac{\varepsilon\mu}{4\pi}(\boldsymbol{E}\times\boldsymbol{H})_i = \frac{\varepsilon\mu}{c}S_i. \tag{3.5}$$

In view of the relationship

$$S_M{}^{i0} = cg_M{}^i,$$

we obtain, therefore, the expression

$$\boldsymbol{g}_M = \frac{\varepsilon\mu}{c^2}\boldsymbol{S} = \frac{n^2}{c^2}\boldsymbol{S} \tag{3.5a}$$

for the momentum density g_M. This, however, would appear to contradict the Planck principle of inertia

$$\boldsymbol{S} = c^2\boldsymbol{g},$$

associating a momentum density $\boldsymbol{g}$ with each energy flux $\boldsymbol{S}$.
Similarly, a contradiction arises also in connection with the principle of equivalence of energy and mass, $E = mc^2$. This is not surprising, as the latter principle is just an integrated form of the Planck principle of inertia [2]. The asymmetry of tensor $S_M{}^{\mu\nu}$ also must give rise to uncompensated mechanical torques which, clearly, are entirely unphysical [2].
The above deficiencies in the Minkowski tensor, all arising from the asymmetric character of that tensor, are rather disturbing and, in an attempt to overcome them, a different form for the energy-momentum tensor (it will be denoted here by $S_A{}^{\mu\nu}$) was proposed by Abraham [52]

$$S_A{}^{\mu\nu} = \frac{1}{4\pi}\left[g^{\mu\alpha}F_{\alpha\lambda}H^{\lambda\nu} + \frac{1}{4}g^{\mu\nu}F_{\alpha\beta}H^{\alpha\beta} + (\varepsilon\mu - 1)\,g^{\mu\alpha}\Omega_\alpha v^\nu\right], \tag{3.6}$$

which, evidently, differs from $S_M{}^{\mu\nu}$ by the addition of the last term, in which

$$g^{\mu\alpha}\Omega_\alpha = F_\beta v_\varkappa(v^\mu H^{\beta\varkappa} + v^\varkappa H^{\beta\mu}),$$

with $F_\alpha = F_{\alpha\beta}v^\beta$, and where v^λ is the four-velocity of the medium. The components of this new tensor are

$$S_A{}^{ik} = \frac{1}{8\pi}\left[(E_iD_k + E_kD_i) + (H_iB_k + H_kB_i) - \delta^{ik}(\boldsymbol{E}\cdot\boldsymbol{D} + \boldsymbol{H}\cdot\boldsymbol{B})\right]$$

$$S_A{}^{0i} = S_A{}^{i0} = \frac{1}{4\pi}(\boldsymbol{E}\times\boldsymbol{H})_i \tag{3.6a}$$

$$S_A{}^{00} = \frac{1}{8\pi}(\boldsymbol{E}\cdot\boldsymbol{D} + \boldsymbol{H}\cdot\boldsymbol{B}).$$

The space-time components form a quantity which, again, should be identifiable with the momentum density (g_A) of the field, while satisfying

$$g_A = \frac{\mathbf{S}}{c^2}, \tag{3.7}$$

in accordance with the Planck principle.

Using the prescription (2.42), for the formation of the force density in the field, we then have, on comparing the resulting expressions for the Minkowski and Abraham tensors, the relation

$$\boldsymbol{f}_A = \boldsymbol{f}_M + \frac{n^2 - 1}{c^2} \frac{\partial \mathbf{S}}{\partial t}, \tag{3.8}$$

while

$$f_A{}^0 = f_M{}^0 \tag{3.9}$$

is the corresponding time component. It is clear, from inspection of (3.8), that the Abraham force density differs from the Minkowski one by the very small additive term.

The macroscopic four-force density $f_A{}^\mu$ is to be regarded as a statistical average of its microscopic equivalents. Therefore, the relation $f_A{}^\mu = \partial_\nu S_A{}^{\mu\nu}$ implies that the macroscopic energy-momentum tensor, in analogy with the four-force, is equal to the averaged microscopic version, because the symmetry of the tensor is preserved under the required operation. It appears that it was this reasoning that initially inclined PAULI [2] towards favouring the symmetric tensor.

EINSTEIN and LAUB [21] arrived at an expression for the energy-momentum tensor which was completely different from the two previously discussed. Starting from simple examples, they first calculated the force exerted by electric, or magnetic, fields on a material volume. They then obtained the force due to the electric dipoles. In other words, they assumed that the interaction between matter and an electromagnetic field was mediated through some hypothetical elementary electric and magnetic particles. The observed force density then had to consist of two contributions, viz., a part depending of the velocity of those particles, and a part which did not. Essentially, this separation was to be understood as a splitting of the total observed force into a surface force $(1 - \mu^{-1}) (\boldsymbol{j} \times \boldsymbol{H}_{\text{ext}})/c$ and a volume force $(\boldsymbol{j} \times \boldsymbol{H}_{\text{int}})/c$. Consequently, the corresponding energy-momentum tensor, was written as

$$\begin{aligned} S_E{}^{ik} &= \frac{1}{4\pi}\left[E_i D_k + H_i B_k - \frac{1}{2}\,\delta^{ik}(\boldsymbol{E}^2 + \boldsymbol{H}^2)\right] \\ S_E{}^{0k} - S_E{}^{k0} &= \frac{1}{4\pi}\,(\boldsymbol{E} \times \boldsymbol{H})_k. \end{aligned} \tag{3.10}$$

It is to be noted that this tensor was not completely determined, as its time-time component was not specified.

Another proposal regarding the energy-momentum tensor was put forward by BECK [22]. In order to reconcile the validity of the velocity addition theorem (which agreed with the conclusion of Minkowski), on one hand, with the Planck principle of inertia (satisfied by the Abraham tensor), on the other, he introduced what he called the radiation tensor $S_B{}^{\mu\nu}$ of the form

$$S_B{}^{\mu\nu} = S_M{}^{\mu\nu} + U^{\mu\nu}$$

where $U^{\mu\nu}$ represents the matter tensor. This radiation tensor may be put in a covariant form

$$S_B{}^{\mu\nu} = \frac{1}{4\pi} \frac{v_\varrho v_\sigma}{v_\alpha v_\beta} \frac{S_M{}^{\varrho\mu} S_M{}^{\sigma\nu}}{S_M{}^{\alpha\beta}}. \tag{3.11}$$

It turned out, however, that, although the tensor $S_B^{\mu\nu}$ was symmetric, some conclusions drawn from it by that author were incorrect [10].
The question was further examined by MARX and GYÖRGYI [23]. On considering the field and the dielectric as forming a closed system, they also found (similarly to Beck) that the field energy-momentum tensor, as given by Abraham, had to be supplemented by a matter tensor. The covariant form of the Marx and Györgyi tensor was

$$S_s^{\mu\nu} = \frac{1}{\varepsilon\mu} \left\{ \frac{1}{4\pi} \left[g^{\mu\alpha} F_{\alpha\lambda} H^{\lambda\nu} + \frac{1}{4} g^{\mu\nu} F_{\alpha\beta} H^{\alpha\beta} + (\varepsilon\mu - 1)\, U_s^{\mu\nu} \right] \right\}, \tag{3.12}$$

where

$$U_s^{\mu\nu} = v^\mu H^{\nu\alpha} F_\alpha + \frac{1}{2} v^\mu v^\nu \left[(\varepsilon\mu - 1)\, F_\alpha H^\alpha + \frac{1}{2} F_{\alpha\beta} H^{\alpha\beta} \right].$$

In the rest frame of the medium, the components of the radiation tensor (3.12) are

$$S_s^{ik} = \frac{1}{n^2} S_A^{ik}, \qquad S_s^{0\alpha} = S_s^{\alpha 0} = S_A^{\alpha 0}. \tag{3.13}$$

Although the tensors of Beck (3.11) and of MARX and GYÖRGYI (3.12) appeared to be quite different, it was shown by SCHÖPF [53] that, at least in the case of a pure radiation field, they were, in fact, identical. Nevertheless, while BECK [22] advanced arguments in support of the Minkowski tensor being taken together with additive terms due to matter, MARX and GYÖRGYI [23] accepted the Abraham version when further supplemented by a contribution again arising from presence of matter.
In an extensive analysis of the question of the energy-momentum tensor, DE GROOT and SUTTORP [24] adopted an approach which was purely microscopic, i.e., one which was related to the behaviour of point particles. According to their thesis, this was the only rigorous way to arrive at the solution of the problem. Thus, they began with a microscopic force law for a constituent particle. The total electromagnetic field acting on such a particle is, of course, equal to the external field, to which the system is subjected, modified by contributions of the neighbouring particles. Put otherwise, the field may be seen as made up of two parts. First, there is the external field together with the addition of the field due to the surrounding particles — this field varies only very little over the atomic distances. Second, there is the intra-atomic field, due to the constituent particles of the atom. Obviously, that field varies quite drastically over the atomic dimensions. Utilizing this concept, de Groot and Suttorp defined an atomic field energy-momentum tensor $t_f^{\alpha\beta}$. Then, from the behaviour of a particle in that field, the atomic material energy-momentum tensor $t_m^{\alpha\beta}$ was obtained. They then assumed that the atomic conservation laws had the form

$$\partial_\beta t^{\alpha\beta} = 0, \qquad t^{\alpha\beta} = t_f^{\alpha\beta} + t_m^{\alpha\beta}. \tag{3.14}$$

The tensor $t^{\alpha\beta}$ thus obtained was, naturally, symmetric, hence implying the law of conservation of angular momentum. If then $m^{\alpha\beta\gamma}$ is the (density) angular momentum tensor, we have

$$\partial_\gamma m^{\alpha\beta\gamma} = 0, \qquad m^{\alpha\beta\gamma} = x^\alpha t^{\beta\gamma} - x^\beta t^{\alpha\gamma}. \tag{3.15}$$

In order to change from densities to observable quantities, a space integration at constant time is applied, thus, in turn, leading to a conserved quantity.
On performing a statistical averaging on such microscopic systems, de GROOT and SUTTORP [24] obtained a macroscopic conservation law of the form

$$\partial_\beta S_G^{\alpha\beta} = 0, \qquad S_G^{\alpha\beta} = \langle t^{\alpha\beta} \rangle. \tag{3.16}$$

The total macroscopic energy-momentum tensor $S_G^{\alpha\beta}$ is symmetric and, in analogy with its microscopic counterpart $t^{\alpha\beta}$, consists of the sum of field and material tensors. Although the tensor $S_G^{\alpha\beta}$ is an average of $t^{\alpha\beta}$, this does not apply to the tensors making up the $S_G^{\alpha\beta}$. In other words, the macroscopic field (matter) tensor is not the mean of the microscopic field (matter) equivalents. Thus, according to the above workers, the averaging procedure applies only to the total energy-momentum tensor.

The symmetry of $S_G^{\mu\nu}$, fulfilled in isotropic media, breaks down when the system is anisotropic. This is clearly seen when the tensor is written in the three-dimensional formalism

$$S_G^{ik} = \frac{1}{4\pi}\left[E_i D_k + H_i B_k - \delta^{ik}\left(\frac{1}{2}E^2 + \frac{1}{2}B^2 - \boldsymbol{M}\cdot\boldsymbol{B}\right)\right]$$

$$S_G^{0i} = S_G^{i0} = \frac{1}{4\pi}(\boldsymbol{E}\times\boldsymbol{H})_i, \qquad S_G^{00} = \frac{1}{8\pi}(E^2 + B^2), \tag{3.17}$$

where $\boldsymbol{M} = \boldsymbol{B} - \boldsymbol{H}$ is the magnetization vector.

In polar substances, where $\boldsymbol{M} = 0$, the tensor $S_G^{\mu\nu}$ coincides with the Einstein and Laub tensor $S_E^{\mu\nu}$, as seen on comparing (3.17) with (3.10).

DE GROOT and SUTTORP [24] proposed also an alternative energy-momentum tensor. That tensor ($S_D^{\mu\nu}$) was defined as the difference between the total energy-momentum tensor in the presence of the electromagnetic field and the corresponding tensor, when such a field was absent. The last of these was, of course, identical with the material tensor. Assuming linear constitutive relations, the new tensor had the following components in the rest frame

$$S_D^{ik} = \frac{1}{4\pi}\left[E_i D_k + H_i B_k - \frac{1}{2}\delta^{ik}\left(\boldsymbol{E}\cdot\boldsymbol{D} + \boldsymbol{H}\cdot\boldsymbol{B} + E^2 V\frac{\partial\varkappa}{\partial V} + B^2 v\frac{\partial\chi}{\partial V}\right)\right]$$

$$S_D^{0i} = S_D^{i0} = \frac{1}{4}(\boldsymbol{E}\times\boldsymbol{H})_i, \tag{3.18}$$

$$S_D^{00} = \frac{1}{8\pi}\left[\boldsymbol{E}\cdot\boldsymbol{D} + \boldsymbol{H}\cdot\boldsymbol{B} + E^2 T\frac{\partial\varkappa}{\partial T} + B^2 T\frac{\partial\chi}{\partial T}\right],$$

where V stands for the specific volume, T for the temperature, and the electric $\varkappa$ and magnetic susceptibilities χ are defined as $\boldsymbol{P} = \varkappa\boldsymbol{E}$ and $\boldsymbol{M} = \chi\boldsymbol{B}$. If it is assumed that the derivatives of the material constants occurring in (3.18) vanish and the medium is isotropic, then it is immediately seen, on comparing (3.18) with (3.6a), that $S_D^{\mu\nu}$ is identical with $S_A^{\mu\nu}$.

All of the above described energy-momentum tensors occur frequently in literature. In the section below, the just outlined forms of these tensors will be examined and the arguments put forward to support them will be discussed.

2. Theoretical considerations

We shall now examine the applicability of the various tensors to the description of different physical phenomena. Also, as conclusions reached in numerous papers, published in many different journals, are quite contradictory, it will be instructive to investigate, at this stage, the different approaches employed in an attempt to resolve the present dilemma.

Proponents of a symmetrical field energy-momentum tensor often refer to a simple thought experiment devised by BALAZS [9]. Although stemming from a rather naive argument, it is, nevertheless, based on sound physical principles of classical physics and it leads to correct results.

Consider, then, two identical and parallel channels with no external forces acting on them. Let each of these channels contain an identical medium. A beam of radiation can be then chosen in such a way that it passes through the given medium in the first channel, but by-passes it in the second. (The selected paths are assumed, for simplicity, to be parallel.) Now, it is obvious that the speed of energy propagation in the medium is smaller than that in the empty space. Hence, the time a photon takes to travel a certain distance is different in the two channels. Put otherwise, a photon (and so the mass associated with its energy) in the first channel requires a longer time to cover that length than does a photon in the second channel. Suppose, the medium does not move while a wave is travelling through it. Then, the displacement of the centre of mass of each channel should be different, in spite of the fact that no external forces are present. This, however, contradicts the velocity invariance. Therefore, the medium, during the passage of an electromagnetic wave through it, must have been displaced by the same amount. It means, in turn, that the medium has a certain momentum and, due to the conservation of the momentum of the total system, value of that momentum in the medium is different from that in vacuo. Put otherwise, as the displacement velocity of the centre of mass of the system is the same in both cases, its position of the first channel is shifted further from the initial one. This, therefore, implies that the medium itself must also move in the same direction. Hence, it may be said, the momentum is transferred to the medium in the direction of motion. This, however, would not be possible had the electromagnetic field momentum in a medium been that given by the Minkowski expression (3.5). (It has been tacitly assumed here that the refractive index of the medium is greater than unity.)

The above may be expressed quantitatively, as follows. If x_c is the x-coordinate of the centre of mass of the system, W — the mass of the medium, and w — the relativistic mass equivalent of the energy carried by the wave, then the velocity of the centre of mass of the wave-medium system is

$$\partial_t x_c = (W\,\partial_t x_1 + w\,\partial_t x_2)/(W + w), \tag{3.19}$$

where x_1 and x_2 are the coordinates of the centre of mass of the material medium and the wave-field, respectively. If, further, $(\partial_t x_c)_0$ is the value of the velocity of centre of mass before the interaction, then

$$(\partial_t x_c)_0 = wc/(W + w). \tag{3.20}$$

From the momentum balance, it then follows that

$$W\,\partial_t x_1 = (h\nu/c) - p, \tag{3.21}$$

where p denotes the momentum associated with the electromagnetic field inside the medium. Let $(\partial_t x_c)_i$ be the velocity of the centre of mass, while the wave-medium interaction is taking place, i.e., while the wave travels through the medium. Then, combining (3.19) with (3.21) leads to

$$(\partial_t x_c)_i = \big((h\nu/c) - p + (wc/n)\big)/(W + w), \tag{3.22}$$

where $\partial_t x_c = c/n$ is the group speed of the wave motion. That speed, on the assumption of a simple medium, is identical with the energy (and, hence, mass) transport.

Now the principle of invariance of the velocity of the centre of mass requires that

$$(\partial_t x_c)_0 = (\partial_t x_c)_i. \tag{3.23}$$

On inserting the relevant expression into (3.23) and on applying the principle of mass-energy equivalence ($w = h\nu/c^2$), we obtain

$$p = h\nu/cn, \tag{3.24}$$

which is the Abraham expression for the field momentum inside the medium.

The above considerations can be summed up in the following way. The total momentum of a system (radiation plus matter) can be uniquely split into two parts [9], with one part referring to the change in the field momentum and the other describing the increase in the mechanical momentum. It can be then concluded that only the symmetric field energy-momentum tensor simultaneously satisfies the momentum conservation and the centre of mass theorems.

Essentially the same thought experiment was also used by other workers. Thus, e.g., BURT and PEIERLS [11] employed this argument for illustrative purposes. They pointed out, however, that the controversy was not adequately resolved by the above reasoning, as the assumption of a rigid medium was involved — the concept incompatible with the special theory of relativity.

A similar analysis was also put forward by JONES [14]. Utilizing rather simple concepts, he attempted to find out how the momentum of radiation inside a medium was distributed between its electromagnetic and mechanical parts which then led to the conclusion that the total momentum associated with the radiation field should be that given by the Minkowski expression, i.e., $p = nh\nu/c$. If, then, n_g refers to the group refractive index, the fraction $\mathring{p}/n_g$ ($\mathring{p}$ — momentum in vacuo) may be (according to Jones) regarded as residing in the wave and, hence, constitutes the electromagnetic part of the total momentum, while the part $np\big(1 - (1/nn_g)\big)$ (n is the conventional phase refractive index) is, according to him, to be associated with the mechanical momentum present in the body. It should be noticed that, in order to arrive at this conclusion, Jones introduces an extra (group) refractive index, n_g, thus, complicating the issue. Besides, in the experiment of JONES and LESLIE [26], one of the conclusions was that the refractive index should be given by the ratio of the phase, rather than group, speeds. The above analysis by Jones seems, therefore, rather unsatisfactory.

A more profound approach is that of HAUS [16]. In order to arrive at the expression for the electromagnetic momentum density in a linearly polarizable medium, he separates the total system into a part describing the motion of a mechanical continuum and another part corresponding to electromagnetic and thermodynamic properties. Then, the requirement for the mechanical momentum conservation of the continuum takes the form

$$\partial_t(\varrho \boldsymbol{v}) + \nabla \cdot (\varrho \boldsymbol{v}\boldsymbol{v}) = f_k,$$

while that for the other part gives

$$\partial_t g_k + \nabla \cdot S_{ik} = -f_k.$$

Here, ϱ is the mass density, v — the velocity, S_{ik} — the stress tensor, and g_k — the momentum density of the remaining, non-mechanical components. Although the separation of the system into kinetic and electromagnetic parts is, essentially, embedded in work of BALAZS [9], the conclusion reached by Haus is slightly modified by some additive terms. He argues that, as the ratiation enters a dielectric, an interaction between field and matter takes place. Thus, the effect of matter upon the field is reflected in the electromagnetic constitutive relations. The force with which the field acts on the medium was calculated by PENFIELD and HAUS [16]. It will be sufficient to mention here that the volume integration of that force results in

$$\int f_k \, dV = -\partial_t \int (\boldsymbol{E} \times \boldsymbol{H}) \, dV/c^2. \tag{3.25}$$

From this, it directly follows that the electromagnetic momentum density in a non-dispersive linear medium is $(\boldsymbol{E} \times \boldsymbol{H})/c^2$.
The momentum exchanged between the field and the medium, is of course, equal to difference between the field momentum before and after the interaction has taken place. Haus then argued that the momentum density possessed by the electromagnetic wave packet travelling in a medium need not be necessarily given by $(\boldsymbol{E} \times \boldsymbol{H})/c^2$ once the said exchange is taken into account.
Furthermore, it is shown that in the case of normal incidence at a sharp boundary between the vacuum and the medium, there are no singularities in the force density f_k. Consequently, there are no surface forces. This implies that all of the momentum is then imparted to the bulk of the medium. Hence, he concludes that all of the momentum is contained in the incident, reflected, and refracted wave, as none is left on the surface. These considerations apply to a medium with negligible electro- and magneto-striction. Thus, finally, it appears as if neither g_A, given by (3.7), nor g_M, from (3.5), could represent the wave momentum. According to Haus, the correct value for the momentum is given by (3.7) supplemented by terms describing a mechanical momentum, attributable to a medium. In addition, the energy flux density must also be increased by a contribution from the medium.
The conclusions reached by Haus, essentially, agree with results of Shockley [*15*]. That author, using an ingenious model, succeeds in interpreting the connection between the macroscopic laws and the structure of the medium. The electromagnetic theory is, first of all, expanded to include the magnetic equivalents of electric charges thus, naturally, leading to the magnetic analogue of the Lorentz force [*54*]. Consequently, the hidden linear momentum is introduced. Its application to the relativistic energy-momentum tensor is discussed [*15*] and, in conjunction with the centre of mass theorem, points to the conclusion that, in stationary matter, the electromagnetic momentum density is correctly given by the Abraham expression. This is supported by the experiment of James [*55*] whose results definitely reject the Minkowski form.
The physical reality of the magnetic dual of the Lorentz force was previously discussed by Costa de Beauregard [*56*] in his attempt at analyzing the Goillot [*57*] experiment. Although the theoretically deduced effect was qualitatively similar to that observed in the experiment, its magnitude was much smaller. In Costa de Beauregard's thought experiment, a ferromagnetic tubular test body and an axial wire with a constant density of charge are considered. Calculations of the force acting on the test body [*58*], assuming conservation of linear momentum, lead to the Minkowski value of the momentum density. A more complete interpretation of the driving force is given in [*59*].
When investigating the so called Weyssenhof behaviour of a spinning particle, which is, essentially, a classical analogue of the Zitterbewegung of the Dirac electron, Costa de Beauregard [*60*] concludes that the non-collinearity of the velocity and momentum of such particles requires an asymmetric energy-momentum tensor. A similar result was also obtained by Papapetrou [*61*]. Hence, it is easily found [*60*] that the four-momentum of photons in a medium should then be space-like, thus, presumably, resulting in a tachyonic type of behaviour.
Another argument in support of the Abraham tensor is supplied by Györgyi [*62*], who claims that the crucial point in the dispute is the Lorentz force acting on polarization currents. He, then, shows that the Abraham theory gives a correct description of, e.g., the Cherenkov radiation.
It is realized [*10*] that, according to Abraham, the momentum transferred by the photon is $h\nu/nc$ and that transferred to the medium is $(n^2 - 1)\, h\nu/nc$, while the Minkowski ten-

sor gives the value of $nh\nu/c$ for the momentum exchange with the photon and no exchange with the medium. The analysis of the Cherenkov radiation by AGUDIN [63] showed, however, that only the Abraham tensor allows the centre of mass theorem to be satisfied.

In most of the above mentioned works, the centre of mass theorem played an important role. Some other workers, e.g., SKOBEL'TSYN [10], BREVIK [18], while realizing that satisfying the theorem alone was not sufficient to establish the complete solution [17], nevertheless, found it useful, at least, for instructive purposes.

In the work of SKOBEL'TSYN [10], another formulation of the special theory of relativity was used, viz., one employing the imaginary time component $x_4 = ix_0$. His paper analyzed the problem using simple models in a phenomenological picture. In order to simplify the task, it assumed that the medium was non-magnetic and that the wave in question was plane progressive. Further, it assumed the direction of propagation to be parallel to the x-axis. Consequently, the y- and z-components of the energy-momentum tensor vanish, thus reducing the problem to that in a two-dimensional space-time. In the special theory of relativity, the transformation properties of a tensor are expressed as

$$T^{\alpha\beta'} = \alpha_\gamma{}^\alpha \alpha_\delta{}^\beta T^{\gamma\delta}, \tag{3.26}$$

where the primed quantities refer to the moving inertial frame and α's represent the transformation matrices, i.e., $\alpha_\alpha{}^\gamma \alpha_{\gamma\beta} = \delta_{\alpha\beta}$, which can be written as

$$\alpha_{\alpha\beta} = \begin{pmatrix} \gamma & i\beta\gamma \\ -i\beta\gamma & \gamma \end{pmatrix}; \qquad \gamma = \left(1 - \frac{v^2}{c^2}\right)^{-1/2}. \tag{3.27}$$

According to (3.6a), the Abraham tensor in the rest frame takes the form [10]

$$S_A{}^{\alpha\beta} = \begin{pmatrix} u & iu/n \\ iu/n & -u \end{pmatrix}, \tag{3.28}$$

where u is the energy density. Similarly, in view of (3.3), (3.4), and (3.5), the Minkowski tensor in the rest frame is written as

$$S_M{}^{\alpha\beta} = \begin{pmatrix} u & inu \\ iu/n & -u \end{pmatrix}. \tag{3.29}$$

Transforming $S_A{}^{\alpha\beta}$ and $S_M{}^{\alpha\beta}$, according to (3.26) and using (3.27) leads to

$$S_A{}^{\alpha\beta'} = \frac{u_0}{n}\gamma^2 \begin{pmatrix} n(1+\beta^2) - 2\beta & i(1 - 2\beta n + \beta^2) \\ i(1 - 2\beta n + \beta^2) & -[n(1+\beta^2) - 2\beta] \end{pmatrix} \tag{3.28a}$$

and

$$S_M{}^{\alpha\beta'} = \frac{u_0}{n}\gamma^2 \begin{pmatrix} (n-\beta)(1-n\beta) & i(n-\beta)^2 \\ i(1-n\beta)^2 & -(n-\beta)(1-n\beta) \end{pmatrix}. \tag{3.29a}$$

These two matrices then describe the corresponding two forms for the energy-momentum tensor in a moving inertial frame. These expressions will now be compared in model situation describing the limiting behaviour of matter, viz., the cases of a rigid body and of a rarefied medium.

In the first of these (the case of rigid body) the total mechanical tensor is represented by the stress tensor, whereas in the latter case it is represented by the product of the momentum density and the velocity of the medium. Thus, in the former case, the total mechanical energy-momentum tensor can be written [10] as

$$U^{\alpha\beta} = \begin{pmatrix} -(n^2-1)\,u/n^2 & 0 \\ 0 & -\varrho c^2 \end{pmatrix}, \tag{3.30}$$

where ϱ is the mass density in the rest frame. Of course, in the absence of the field, all components of $U^{\alpha\beta}$ other than U^{44} vanish. Thus, that part of the mechanical tensor which depends on the electromagnetic field is

$$\begin{pmatrix} -(n^2-1)\,u/n^2 & 0 \\ 0 & 0 \end{pmatrix}. \tag{3.31}$$

The component $U^{11} = (n^2 - 1)\,u/n^2$ is often referred to as the radiation reaction force [10]. Transforming (3.31) into the moving frame, using (3.26) and (3.27), gives the following expression for the tensor, $P_1{}^{\alpha\beta}$, which represents the response of the medium as observed from the moving frame

$$P_1{}^{\alpha\beta} = \gamma^2 p_0 \begin{pmatrix} 1 & -i\beta \\ -i\beta & -\beta^2 \end{pmatrix}, \tag{3.32}$$

where $p_0 = -(n^2 - 1)\,u/n$.
Now, the sum of $P_1{}^{\alpha\beta}$ and $S_A{}^{\alpha\beta}$, as given by (3.32) and (3.28), respectively, is the total tensor for the system

$$T_1{}^{\alpha\beta} = \frac{u_0}{n^2}\gamma^2 \begin{pmatrix} (n\beta-1)^2 & i(n-\beta)(1-\beta n) \\ i(n-\beta)(1-\beta n) & -(n-\beta)^2 \end{pmatrix}. \tag{3.33}$$

Let us next consider, together with SKOBEL'TSYN [10], the second limiting case, viz., that of a "dust-like" medium. In that case, all stresses vanish and the corresponding tensor is then a product of the momentum density and the velocity of the medium. This assumes that the mass of each of the constituent particles is approaching infinity, while their corresponding number density is rather small. Consequently, in this approximation, the refractive index may be taken to be independent of the intensity of radiation (as the particle mass is very large). In addition, the radiation is not reflected at the incidence. When an electromagnetic beam enters the medium, the surface of the latter moves towards the source with a mechanical momentum [10]

$$p_2 = (n - 1)\,h\nu/c. \tag{3.34}$$

It should be remarked that the same expression is obtained from both the Minkowski and the Abraham tensors.
Applying now the conservation equations (3.19), (3.21), and (3.23), the momentum p_1 of the material component of the system becomes

$$p_1 = (n - 1)\,h\nu/cn. \tag{3.35}$$

This momentum, however, is equal to the difference between the medium and surface momenta. Thus, addition of (3.34) and (3.35) leads to the expression

$$p_3 = (n^2 - 1)\,h\nu/cn \tag{3.36}$$

for the momentum of the medium.
Furthermore, if the particle number density in the absence of the field is ϱ_0, then, at some later time t, it becomes

$$\varrho(t) = \varrho_0 + \Delta\varrho. \tag{3.37}$$

On using the continuity equation, we have [10]

$$\Delta\varrho = (n^2 - 1)\, u_0(t)/c^2. \tag{3.38}$$

The mechanical energy-momentum tensor may then be obtained in the form [10]

$$U^{\alpha\beta} = \begin{pmatrix} \beta_0^2\varrho_0c^2 & i\dfrac{n^2-1}{n}u_0(t) \\ i\dfrac{n^2-1}{n}u_0(t) & -\varrho_0c^2 - (n^2-1)\,u_0(t) \end{pmatrix}. \tag{3.39}$$

On subtracting the value of $U^{\alpha\beta}$ in the absence of the field from (3.39), we obtain an expression for that part of the tensor which is related to the field

$$P_2^{\alpha\beta} = \begin{pmatrix} \beta_0^2\varrho_0c^2 & \dfrac{n^2-1}{n}u_0(t) \\ i\dfrac{n^2-1}{n}u_0(t) & -(n^2-1)\,u_0(t) \end{pmatrix}, \tag{3.40}$$

where $\beta_0c = (n^2 - 1)\, u_0/\varrho_0cn$. In the limit of infinitesimally small number density, the above tensor reduces to

$$P_2^{\alpha\beta} = \frac{n^2-1}{n}u_0(t)\begin{pmatrix} 0 & i \\ i & -n \end{pmatrix}. \tag{3.41}$$

A subsequent transformation to the moving inertial frame yields

$$P_2^{\alpha\beta} = \frac{n^2-1}{n}u_0(t)\,\gamma^2\begin{pmatrix} n\beta^2 - 2\beta & i(1+\beta^2-n\beta) \\ i(1+\beta^2-n\beta) & -(n-2\beta) \end{pmatrix}. \tag{3.42}$$

The energy-momentum tensor corresponding to the whole system is then constructed, in analogy to above, resulting in

$$T_2^{\alpha\beta} = u_0\gamma^2\begin{pmatrix} (n\beta-1)^2 & i(n-\beta)(1-n\beta) \\ i(n-\beta)(1-n\beta) & -(n-\beta)^2 \end{pmatrix}. \tag{3.43}$$

Together with SKOBEL'TSYN [10, 35], we can now compare the two expressions, (3.33) and (3.43), for the total energy-momentum tensor applicable in each limiting case. It is interesting to note that those forms differ from each other only by the factor $1/n^2$.

We can now see the reason for considering two limiting cases of media, in which the interaction with field is to be investigated. For, clearly, the splitting of the total tensor of the system into contributions corresponding to the field and the matter, is not simple [17, 18, 65, 66]. Skobel'tsyn, therefore, in an attempt to overcome this difficulty, considers two extreme model situations, viz., these of rigid and "dust-like" media. This makes it possible to divide up the total energy-momentum tensor uniquely into field and matter parts, provided that the field energy-momentum tensor satisfies

$$\partial_\beta S^{\alpha\beta} = -f^\alpha, \tag{3.44}$$

(in accord with the discussion in Chapter II).

SKOBEL'TSYN, further, claims [10] that the momentum density S^{14} is obtained from the invariance of the velocity of centre of mass together with the momentum conservation law, while to determine the component S^{41}, one uses (3.44), (in the form $\partial_\beta S^{4\beta} = 0$). As the left hand side of the last equation expresses the work done by the field, that equation then describes the continuity of the energy flux (proportional to the Poynting vector). Consequently, $S^{41} = S^{14}$, thus implying the symmetry of $S^{\alpha\beta}$. From the requirement that the energy-momentum tensor of the total system be symmetric, it follows that also the mechanical tensor must be symmetric. Hence, Skobel'tsyn concludes that the correct expression for the field is that given by Abraham.

It should be noted that, initially, von Laue also [67] accepted the reasoning of Abraham. However, in a later article [6], he developed an argument which (so he believed) rejected this tensor and replaced it by the Minkowski one. Calculations of Scheye [68] even further supported this idea. The following considerations led to this change.

It is well known that, in general, when plane electromagnetic waves propagate through a medium, their phase speed is different from their group speed. Further, if the absorption in the medium is low, the latter speed may be identified with the speed of energy propagation. Von Laue [6] required the group velocity to satisfy the velocity superposition theorem due to Einstein, and proceeded to show that only the Minkowski tensor fulfilled this condition. Moller [7] developed a more general version of the same argument. Thus, if $S^{\mu\nu}$ is the electromagnetic energy-momentum tensor, then the velocity of the energy propagation is defined by $v_g{}^i = icS^{4i}/S^{44}$. Moller finds the condition for this velocity to transform like a particle velocity to be

$$R^{\mu\nu} = 0, \tag{3.45}$$

where $R^{\mu\nu} = S^{\mu\nu} + 1/c^2\, S^{\mu\alpha}v_\alpha v^\nu$ and v_α represents the four-velocity. He then shows that only the Minkowski tensor satisfies (3.45) and, thus, concludes that this tensor must give an adequate description of electromagnetic phenomena in matter.

Skobel'tsyn [10], nevertheless, doubts the validity of this criterion.

Next, using the thermodynamics of irreversible processes, Tang and Meixner [69] analyzed the influence of electromagnetic waves on a simple fluid. The total energy-momentum tensor obtained by them consisted of three contributions: a constant field term, an oscillating field term, and a constant matter term. Noting that, in the absence of the field, the velocity of the energy transport does not satisfy the addition of velocities theorem, they split the total energy flow in the separate parts and then examine the von Laue und Moller transformation criterion for these parts. They concluded that Moller's treatment was incomplete and suggested that the Abraham and Minkowski tensors lead to the same result, provided the matter contribution is properly accounted for.

De Groot and Suttorp [24], however, claimed that what Tang and Meixner [69] did, in fact, disproved the very validity of the von Laue criterion. Recently, this last statement was strongly criticized by Brevik [18].

Let us concentrate now on static fields inside a medium, hoping to obtain at least some correct results. We adopt a variational approach of Brevik and assume that the field energy is known. A similar method as also used by Landau and Lifschitz [8].

The free energy of an electrostatic field in a medium is defined, in the usual way [8], by

$$\mathcal{F} = \frac{1}{8\pi} \int \boldsymbol{E} \cdot \boldsymbol{D}\, dV. \tag{3.46}$$

Now, the rate of change of the field free energy may be interpreted as equal to mechanical work $\int \boldsymbol{f} \cdot \boldsymbol{v}\, dV$ (v — velocity, f — force density). One thus obtains the expression for the force density

$$\boldsymbol{f}_A = \varrho \boldsymbol{E} - \frac{1}{8\pi} E_i E_k \nabla \varepsilon_{ik} + \frac{1}{8\pi}\, \partial_k(\boldsymbol{E} D_k - E_k \boldsymbol{D}). \tag{3.47}$$

As, further, the divergence of the stress tensor is equal to the force density, i.e., $f_A{}^i = -\partial_k S_A{}^{ik}$, so (3.47) is compatible with a tensor of the form

$$S_A{}^{ik} = \frac{1}{8\pi}\left[E_i D_k + E_k D_i\right] - \frac{1}{8\pi}\delta^{ik}\boldsymbol{E}\cdot\boldsymbol{D}. \tag{3.48}$$

Comparing it with (3.6a) shows (3.48) to be the same as the expression for the Abraham tensor in the static limit. As this result follows directly from the variational principle, the validity of the tensor is thus established [*8*, *18*].

Now, the crucial point in Brevik's argument consists in the proper accounting for the surrounding medium. For this purpose, consider the existence of torques in a volume element even while the external force is absent. This can happen when there is a distribution of electric dipoles in a medium, describable by the polarization vector $\boldsymbol{P} \doteq \boldsymbol{D} - \boldsymbol{E}$. Then, given the torque, there is an additional amount of work done. In this model, then, the torque density $\boldsymbol{\tau} = \boldsymbol{D}\times\boldsymbol{E}$ represents an extra contribution to the total mechanical work. Putting the rate of change of the free energy equal to that work done per unit time, leads to the Minkowski result for the force density

$$\boldsymbol{f}_M = \varrho\boldsymbol{E} - \frac{1}{8\pi}E_i E_k \nabla\varepsilon_{ik}, \tag{3.49}$$

while the stress tensor takes on the form

$$S_M{}^{ik} = \frac{1}{4\pi}\left[E_i D_k - \frac{1}{2}\delta^{ik}\boldsymbol{E}\cdot\boldsymbol{D}\right], \tag{3.50}$$

which is that given by Minkowski for the static field (cf. (3.2)). This, of course, does not prove the validity of the Minkowski force, $f_M{}^\mu = -\partial_\nu S_M{}^{\mu\nu}$, as its existence, at least in the static case considered here, depends upon the assumption of the body torques. Nevertheless, according to BREVIK [*18*], this supposition is quite natural.

Yet another distinction between the two tensors may be drawn. Thus, if the balance equation is recast in terms of the rotational angle (instead of velocity), then it can be satisfied only by the Minkowski tensor. However, when the quantity, which vanishes after integration, has been added to it, the equality is fulfilled for the Abraham tensor as well. This means that the Abraham density already includes the effect of torques, while, in the Minkowski case it is represented by the asymmetry of the tensor. This is, essentially, one of the main conclusions of BREVIK's [*18*] paper.

Let us now examine the behaviour of the other proposed tensor forms in the time-independent field. Comparison of the Einstein-Laub tensor (3.10) with the first de Groot-Suttorp expression (3.17) shows that, in this (static) limit, both tensors are equal. Hence, it is quite sufficient to consider only one proposal, say, the former one. Thus, the calculation of the force density from the usual Maxwell equations then leads to the expression for the so called Kelvin force density

$$\boldsymbol{f}_E = \varrho\boldsymbol{E} + (\boldsymbol{P}\cdot\nabla)\,\boldsymbol{E}. \tag{3.51}$$

This differs from the corresponding Abraham (3.47) and Minkowski (3.49) results, but is equal to that predicted by LIVENS [*70*].

It is necessary, therefore, first to ascertain which expression is correct. A relatively simple possibility for doing this differentiation occurs in the case of a non-polar liquid dielectric. In that case, it is necessary to include the effect of electrostriction [*51*]. Then, on applying the above variational procedure, we are led to the Helmholtz force

$$f_H = \varrho \boldsymbol{E} - \frac{1}{8\pi} E^2 \nabla \varepsilon + \frac{1}{8\pi} \nabla \left(E^2 \varrho_m \frac{d\varepsilon}{d\varrho_m} \right), \tag{3.52}$$

where the last term refers to the dependence of permittivity on the mass density ϱ_m. Using the Clausius-Mossotti relation, $(\varepsilon - 1)/(\varepsilon + 2) = \text{const.}\ \varrho_m$, we then calculate the excess pressure inside a dielectric as [*30*]

$$\Delta p_H = \frac{1}{24\pi} (\varepsilon - 1)(\varepsilon + 2) E^2. \tag{3.53}$$

On the other hand, the excess pressure connected with (3.51) is

$$\Delta p_E = \frac{1}{8\pi} (\varepsilon - 1) E^2. \tag{3.54}$$

The experiment to test the above mentioned expression for the excess pressure in the presence of the field has, in fact, been performed by Hakim and Higham [*30*], and will be discussed in the last section of this chapter. It will be sufficient to say here, that it has firmly established the validity of (3.53).

In view of that last finding, it can be asserted, that the Einstein-Laub (and also de Groot-Suttorp) tensors yield results which are (at least for static fields) incompatible with experiment.

Consider now a plane wave travelling in an isotropic and homogeneous medium and follow the arguments of Landau and Lifshitz [*8*] about the construction of the energy-momentum tensor from the macroscopic considerations. Those authors assumed that the total energy density is the sum of the electrostatic and magnetostatic energy densities and that the same also applies to the stress tensor. If the frequency of the incident wave is much lower than the characteristic frequency of electronic vibrations, then, for not too strong fields, the relations between the field intensities and corresponding inductions remain linear. Hence, the above described construction of the energy-momentum tensor is valid.

It is found from experiments that the electromagnetic wave travelling through an insulator does not produce, in the first approximation, any heat, as it undergoes an elastic scattering. This, therefore, means that the time component of the four-force vanishes. Consequently, the continuity equation is $\partial_t u + \nabla \cdot \boldsymbol{S} = 0$ and, from the expression for the energy density, $u = (\boldsymbol{E} \cdot \boldsymbol{D} + \boldsymbol{H} \cdot \boldsymbol{B})/8\pi$, one obtains the energy flux as $\boldsymbol{S} = c(\boldsymbol{E} \times \boldsymbol{H})/4\pi$. Extending the validity of the Planck law of inertia, $\boldsymbol{S} = c^2 \boldsymbol{g}$, the expression for the momentum density becomes $\boldsymbol{g} = (\boldsymbol{E} \times \boldsymbol{H})/c$. Hence, the energy density, energy flux, and momentum density constitute the Abraham tensor, as is seen from (3.6a).

Further, if the mechanical part of the energy-momentum tensor is $u^{\mu\nu}$, then we can write

$$\partial_\nu u^{\mu\nu} = f_A{}^\mu. \tag{3.55}$$

Of course, the electrostrictive and magnetostrictive effects are not considered here. Combining (3.55) with the corresponding relation for the field part of the total tensor, $\partial_\nu S_A{}^{\mu\nu} = -f_A{}^\mu$, allows to interprete $(S_A{}^{\mu\nu} + u^{\mu\nu})$ as the total tensor describing a closed system. This tensor is symmetric and so are its constituents.

For Brevik [*18*], however, the argument does not end there. He believes that, when properly interpreted, the Abraham and Minkowski tensors will be equivalent in the majority of situations. He advocates the symmetry of the total energy-momentum tensor of the whole system but its separation, merely motivated by convenience, into two asymmetrical parts is not well supported and remains rather dubious.

At least in principle, it is inadmissible to equate the Abraham and Minkowski results. For instance, the Abraham force is the real force which may be observed to be acting on charges and on matter present in a volume. Reaction to this force results in a mechanical stress. On the other hand, the meaning of the Minkowski force is rather unclear.
Another paper dealing with the radiation forces in dielectric media, written by GORDON [12], demonstrates that the true momentum density of the electromagnetic field in a non-dispersive medium is, indeed given by the Abraham expression. Gordon adopts a microscopic model and begins with a simple, weakly polarizable medium, such as a gas of heavy atoms. He, further, assumes that the corresponding permittivity only slightly exceeds unity and, therefore, the electromagnetic properties of the medium are completely described by its polarizability α. Now, in the rarified gas, the ponderomotive force is simply the Lorentz force, which can be written as [12]

$$\boldsymbol{f}_1 = \alpha \left[(\boldsymbol{E} \cdot \nabla) \boldsymbol{E} + \frac{1}{c} \frac{\partial \boldsymbol{E}}{\partial t} \times \boldsymbol{H} \right]. \tag{3.56}$$

For a plane wave in a medium, the first term vanishes and, thus, the force density due to N atoms is $\boldsymbol{f} = N\boldsymbol{f}_1$. The time integration of this force gives then the mechanical momentum travelling with the pulse as

$$\boldsymbol{g}_{\text{mech}} = \frac{n-1}{4\pi c} (\boldsymbol{E} \times \boldsymbol{H}). \tag{3.57}$$

From the laws of conservation of energy and momentum, Gordon then obtains

$$n\boldsymbol{g} = \boldsymbol{g} + \boldsymbol{g}_{\text{mech}},$$

thus interpreting $\boldsymbol{g} = \boldsymbol{E} \times \boldsymbol{H}/4\pi c$ as the electromagnetic momentum density. Hence, the possible separation of the total momentum into field and matter parts leads, in this model, to a complete solution. Although he proceeds with generalization of the gaseous state to include fluids and solids, it does not appear that his model will lend itself to an easy generalization, as it basically applies only to media with low polarizations.
A different resolution of the present dilemma is advocated by GINZBURG [19]. Although he considers that no objection can be made against the Abraham tensor, which is more fundamental, and correct in the general case, he, nevertheless, attempts to explain why the quantum mechanical analysis of the problem leads to the accepted conservation laws. Thus, first of all, he derives the laws, using the Hamiltonian method, and proceeds then to compare them with the results of macroscopic electrodynamics. From the point of view of the latter, the conservation laws

$$\partial_k S_A{}^{ik} = f^i; \qquad f^\alpha = f_L{}^\alpha + f_A{}^\alpha; \qquad f^4 = \frac{i}{c} \boldsymbol{j} \cdot \boldsymbol{E} \tag{3.58}$$

and

$$\partial_k S_M{}^{ik} = f_L{}^i; \qquad f_L{}^4 = \frac{i}{c} \boldsymbol{j} \cdot \boldsymbol{E}, \tag{3.59}$$

where $f_L{}^\alpha$ is the Lorentz force density, are identical [19]. They differ only by the separation of the four-force in two different ways. Mathematically speaking, this conclusion is correct. However, from the empirical point of view, this resolution is not satisfactory, as the Abraham force is, in principle, measurable, and Ginzburg is aware of this but, nevertheless, prefers, when convenient, to use the Minkowski tensor. In other words, he emphasizes that, when proper account is taken of the background medium, both tensors will, in most situations, yield the same result.

Elementary considerations in support of the Abraham proposal were also put forward by BURT and PEIERLS [11]. It is accepted that, as the wave propagates through a medium, the motion of atoms of that medium is affected, so that these may acquire some additional momentum. In their discussion of the total momentum, both authors, however, did not consider this mechanical contribution. In other words, their remarks apply to a medium which, effectively, does not disturb propagation of light in it.

In a more recent paper, PEIERLS [13] rectifies this shortcoming. He begins with a discussion of the forces acting on the atoms in a medium. In agreement with the result found previously by Gordon [12], the electromagnetic momentum density, he asserted, is equal to the Abraham value. He arrives at this conclusion by realizing that most refractive media are non-magnetic, so that it is permissible to write $\mu_0 \bar{\boldsymbol{H}}_{\text{micro}} = \boldsymbol{B}$, describing the macroscopic magnetic induction as an average of the microscopic intensity $\boldsymbol{H}_{\text{micro}}$. Assuming the microscopic validity of the Poynting vector and $\bar{\boldsymbol{E}}_{\text{micro}} = \boldsymbol{E}$, leads then directly to the conclusion $\boldsymbol{g}_f = \boldsymbol{E} \times \boldsymbol{H}/c^2$.

In order to determine the total momentum, the value of its other, i.e., mechanical component part must be found. This is obtained by considering the force with which the field acts on an atom. Thus, if $\boldsymbol{d}$ is the atomic electric dipole moment, then this force is

$$\boldsymbol{f} = (\boldsymbol{d} \cdot \nabla)\, \boldsymbol{E}_{\text{eff}} + \dot{\boldsymbol{d}} \times \boldsymbol{B}_{\text{eff}}, \tag{3.60}$$

where E_{eff} and B_{eff} are the effective field strengths at the site of the atom. Peierls then argues that, in a non-magnetic dielectric, $\boldsymbol{B}_{\text{eff}}$ may be replaced by the macroscopic field intensity $\boldsymbol{B}$. In addition, the difference between the effective and external electric field is proportional to the polarization. Making use of the Maxwell equations and calculating the force on all atoms in a given volume, he finally obtains for the density of mechanical momentum the expression

$$\boldsymbol{g}_{\text{mech}} = \frac{n^2 - 1}{2c^2} \left[1 - \frac{n^2 - 1}{5}\right] \boldsymbol{E} \times \boldsymbol{H}. \tag{3.61}$$

Clearly, $\boldsymbol{g}_{\text{mech}}$ represents the momentum density in a medium while an electromagnetic wave is travelling within it. Peierls proceeds then to derive expression for the total momentum density

$$\boldsymbol{g} = \frac{1}{2} \left[\frac{1}{n} + n - \frac{(n^2 - 1)^2}{5n}\right] \boldsymbol{g}_{\text{vac}}, \tag{3.62}$$

where he assumes that the light pulse is wide and short. He obtains (3.62) by adding together the mechanical and the field momentum densities. The factor 1/5 in the last term of (3.62) results from assuming that all dipoles are randomly oriented. That value is found to be valid in the region of optical frequencies, when the motion of atoms during the cycle can be neglected, i.e., when the atoms may be treated as stationary. A value of 1/3 (instead of 1/5) was proposed by LANDAU and LIFSHITZ [8] to apply at lower frequencies, when the atomic motion can be no longer neglected.

One final remark: according to Peierls, neither the Abraham nor the Minkowski expressions give the correct total momentum density of a system. In fact, (3.62) shows that for a weak refraction, i.e., when the value of the refractive index is close to unity, the total momentum is just the average of the Abraham and Minkowski values.

In his more recent paper PEIERLS [71] has shown than the expression (3.62), although originally derived for wide and short light pulses is, in fact, independent of the shape of the pulse as long as its dimensions are much larger than the wavelengths. Further, his investigation leads to forces which depend on the polarization of radiation. Thus, when the planes of incidence and polarization are mutually perpendicular, the pressure exerted on a mirror immersed in a liquid is directly proportional to the refractive index of the latter. However, when the two planes coincide, that pressure is reduced by a factor

$$1 - (1 - n^{-2}) [1 + 3(n^2 - 1)/5] \tan^2 \Theta, \tag{3.63}$$

where Θ is the angle of incidence. Obviously, this result is very different from the Minkowski prediction. An experiment to distinguish between these proposals has been, in fact, performed and will be discussed in the next section.

The Peierls result (3.62) was rederived by WONG and YOUNG [72], who made use of the concept of the surface force. Summing up the forces acting on the dipoles near the surface, while assuming that a dielectric is lossless, they obtain for the surface force density

$$f_s = (n^2 - 1)^2 E^2/5, \tag{3.64}$$

where E refers to the incident wave. It is to be noted that, as this force is due to dipole-dipole interactions, it is found to be proportional to $(n - 1)^2$. On the contrary, both the Abraham and the Minkowski expressions give only the first order dependence.

The derivation [72] of the surface force (3.64) does not take into account the existence of lateral forces, due to the transverse gradient of the external field and which push the medium into the beam. As a consequence, there are two different momentum fluxes present: that of the momentum of the electromagnetic wave, propagating with the group velocity, and that of the momentum of the pressure wave, travelling with the speed of sound. Above authors [72] explicitly consider the former case only.

In the calculations of the radiation pressure on matter, it is generally assumed that the properties of the latter do not change during the interaction. Naturally, this is not the case in reality. Thus, ASKAR'YAN [73] analyses a medium with a variable polarizability α. Such a variation is, in general, associated with changes in the parameters (e.g., shape, volume) characterizing the medium. The resulting change in the dipole energy may greatly affect the field energy and, hence, the corresponding radiation momentum. It is then found that the force connected with the change in polarizability is proportional to $\partial_t\alpha$. Consequently, depending on the sign of $\partial_t\alpha$, the deformation forces may either enhance or diminish effects due to the ordinary radiation pressure.

A purely macroscopic approach is employed by ROBINSON [17]. Although he realizes the complexity of problem at hand, he claims that the solution to it already exists and that it has been discussed by PENFIELD and HAUS (cf. [17]). This is done by extending the principle of virtual work, used originally by Helmholtz. Robinson, for reasons stated in his article, describes that approach in a "digested" form. The macroscopic properties of matter, in his view, are described by dielectric permittivity and magnetic permeability. He emphasizes, however, the importance of the electro- and magneto-strictive terms, arguing that the effect of the electronic structure of matter on the electromagnetic phenomenon cannot be completely described by the constitutive relations.

Manipulating the Maxwell equation, Robinson arrives at an identity when he replaces ϱ and $\boldsymbol{j}$, in $\varrho\boldsymbol{E} + \boldsymbol{j}\times\boldsymbol{B}$, by $\nabla\cdot\boldsymbol{D}$ and $\nabla\times\boldsymbol{H} - \partial_t\boldsymbol{D}$, respectively. In regions where polarization $\boldsymbol{P}$ and magnetization $\boldsymbol{M}$ vector densities vanish. the obtained equation is then interpreted as a momentum balance equation.

Different considerations yield different identities. However, a unique solution cannot be revealed by merely manipulating the field equations. Penfield and Haus consider the possibility of extracting some relevant information from the general conservation laws. The solution for particles moving in vacuo is rather well known and this helps the analysis.

The theory of relativity leads to the velocity invariance of the centre of mass of the system. If the energy of a wave in vacuo is $h\nu$, the mass attributed to this wave is $h\nu/c^2$.

Similar considerations were already discussed in the present work. Robinson generalizes that approach and, although not arriving at the complete solution, confirms that the part $\boldsymbol{E} \times \boldsymbol{H}$ of the total momentum of a field in an isotropic medium does not refer to moving matter.

Consider a nearly monochromatic pulse impinging on a dielectric slab characterized by constants ε and μ. It is assumed that the wave energy is conserved, i.e., the kinetic energy associated with the transferred momentum is much smaller than the rest energy of the medium. Wave incident from vacuum is partially reflected and partially transmitted at the boundary. There is also another term in the momentum balance equation, viz., one describing the momentum imparted to the surface. By writting down the momentum conservation law at incidence, it is found [*17*] that the sum of all surface momenta gives the total momentum transferred to the slab. It is, in fact, equal to the difference between the initial and the final momenta of the waves, i.e., those just before an entry and just after an exit.

Assuming that the total momentum of a wave pulse in a medium is $\alpha h\nu/cn$, and that the transferred momentum is $\beta h\nu/cn$, where the coefficients α and β are related to ε and μ, it is found that [*17*], in view of the center of mass theorem, $\alpha - \beta = 1$. Thus, there remains a part $(\alpha - \beta)\, h\nu/cn = h\nu/cn$ of the total propagating momentum which is not associated with matter and may, therefore, be regarded as corresponding to the field. However, the expression for α is not known and this makes it impossible to attach a definite value to the surface momentum. If it is assumed that this value vanishes at normal incidence, then $\alpha = 1/2(\varepsilon + \mu)$ and the total momentum travelling through a medium is

$$g = \frac{h\nu}{2cn} (\varepsilon + \mu). \tag{3.65}$$

In his earlier treatment [*13*], PEIERLS considered normal incidence and only those media in which the difference between μ and unity could be neglected, to that the refractive index was given by $n = \sqrt{\varepsilon}$. Consequently, for media where $n - 1$ is small, the results (3.62) of Peierls and (3.65) of Robinson are identical.

As already indicated, the Helmholtz method of virtual work was generalized by PENFIELD and HAUS [cf. *17*]. They included moving, dispersive media, by treating the mechanical aspects of the system from a relativistic standpoint. They claimed their so obtained solution to be complete. The essence of their argument is as follows.

The radiation field together with the surrounding medium form a system which may be divided into two parts, according to the way in which the energy and momentum are stored. In one part, this is related to the motion of the medium, while in the other, it is connected to mechanical or electromagnetic stresses. Clearly, these subsystems are not closed and, hence, there must be some mutual interaction. The conservation laws have, therefore, the forms

$$\partial_l T^{kl} + \partial_t g^k = f^k, \qquad \partial_k S^k + \partial_t u = \phi, \tag{3.66}$$

where T^{kl} is the momentum flux tensor, g^k the momentum density, and ϕ refers to the power transfer density. The tensor T^{kl} includes, however, not only the electromagnetic but also the mechanical (e.g., electro- and magneto-strictive) variables. This, in turn, should be understood as pointing to the feasibility of splitting up the total momentum into its field and material contributions. This follows from the fact that the rate of change of the mechanical momentum density must now include not only the spatial derivatives of T^{kl} but also $\partial_t g^k$. The most natural interpretation for the latter term is, then, to identify it with the rate of change of the field momentum density.

Regarding the Abraham-Minkowski controversy, Penfield and Haus concluded that g^k, although not referring to the motion of matter, is important in calculating forces exerted by the field on material particles. Robinson agrees that, in some case, the Minkowski momentum density may enable one to find the total momentum transferred to a medium, however, he advocates [17] the Abraham formula as the rigorously correct one.

A variational approach using the Hamiltonian principle was recently presented by MIKURA [20]. Although it is well known that the Lagrangian density cannot be defined uniquely, its form is suggested intuitively and, then, justified a posteriori. Thus, the material part of the Lagrangian density is

$$\mathcal{L}_m = -(-W_\mu^2)^{1/2} - \varepsilon_i(\varrho_0, s). \tag{3.67}$$

Here, W_μ is a four-vector of mass flow density, i.e., $W_\mu = (\varrho \boldsymbol{v}, i\varrho c)$, $\varrho(x_\mu)$ is the mass density, $v(x_\mu)$ — the velocity of particle. Further, $\varepsilon_i(\varrho_0, s)$ is the non-electromagnetic internal energy and s refers to the entropy.

Another component part of the Lagrangian density is due to the field. It is usually written as (2.16)

$$\mathcal{L}_f = -\frac{1}{16\pi} F_{\mu\nu} F^{\mu\nu}. \tag{3.68}$$

The interaction Lagrangian density is proposed on a basis of rather intuitive arguments. If $P_{\mu\nu}$ and $M_{\mu\nu}$ are polarization and magnetization tensors, respectively, then this density takes the form

$$\mathcal{L}_i = A_\mu j^\mu + \frac{1}{4\alpha} P_{\mu\nu} P^{\mu\nu} - A_\mu \partial_\nu P^{\mu\nu} - \frac{1}{4\beta} M_{\mu\nu} M^{\mu\nu} - A_\mu \partial_\nu M^{\mu\nu *}. \tag{3.69}$$

The first term of (3.69) describes the usual interaction between free charges and the field, while the second and third terms refer to the electric dipole-dipole and the field-dipole interactions. The last two terms are the magnetic analogues of the preceding two terms. Further, $M^{\mu\nu *}$ denotes tensor dual to $M^{\mu\nu}$. The parameters α and β, which depend on ϱ_0 and s, characterize the assumed linear properties of a medium, viz., $\varepsilon = 1 + 4\pi\alpha$ and $\mu = (1 - 4\pi\beta)^{-1}$.

The Hamiltonian principle is next applied to derive the equations of motion. Mikura then shows that the sum of the above postulated component parts of the Lagrangian density leads to the well established formulae for the equations of motion. These are the hydrodynamic equations for the material, the constitutive relations, and the field equations. Clearly, they are mutually independent and, thus, the energy-momentum tensors derived for different subsystems are also mutually independent. In other words, the energy-momentum conservation law for the material part is derivable from the hydrodynamic equations, while the corresponding law for the field is obtained from the field equations. Thus, the conservation law for the complete system is a direct consequence of the conservation law for individual subsystems.

An explicit calculation leads to the following form for the material part of the energy-momentum conservation law

$$\partial_\nu S_m^{\mu\nu} = F^{\mu\nu} \varrho_0 u_\nu - \frac{1}{2} \partial_\mu F^{\gamma\delta} P_{\gamma\delta} - \frac{1}{2} \partial_\mu F^{\gamma\delta *} M_{\gamma\delta}. \tag{3.70}$$

Here, the right hand side describes the effect of the field on the material medium, i.e., it represents, in fact, a force density.

Similarly, the field part of the energy-momentum conservation law is derived as

$$\partial_\nu S_f^{\mu\nu} = -F^{\mu\nu} \varrho_0 u_\nu + \frac{1}{2} \partial_\mu F^{\gamma\delta} P_{\gamma\delta} + \frac{1}{2} \partial_\mu F^{\gamma\delta} M^*_{\gamma\delta}, \tag{3.71}$$

where

$$S_f^{\mu\nu} = \frac{1}{4\pi}\left(F^{\mu\gamma}F_\gamma^{\ \nu} - \frac{1}{4}F_{\gamma\delta}F^{\gamma\delta}\delta^{\mu\nu}\right) + F^{\mu\gamma}P_\gamma^{\ \nu} + F^{\mu\gamma}M_\gamma^{\ \nu*} \quad (3.72)$$

is interpreted as the field energy-momentum tensor.

To find the conservation of the total energy and momentum, MIKURA [20] defines the tensor for the total system, although not uniquely, as the sum of the tensors corresponding to the subsystems and, then, combining (3.70) and (3.71), obtains $\partial_\nu S^{\mu\nu} = 0$.

It is to be noted that the definitions of $S_m^{\mu\nu}$ and $S_f^{\mu\nu}$ are not properly justified on the physical basis. As already indicated, the total tensor may be split into two parts in several different ways, if the formal modifications are made in the conservation laws for the subsystems. Mikura, then, attempts to find a physical criterion which would make this tensor splitting into material and field parts unambiguous. He postulates, therefore, that: 1. the part of the tensor which has a form a of pure material tensor represents the material component is the total tensor; 2. the part identical with the field tensor in vacuo is associated with the field contribution to $S^{\mu\nu}$; 3. tensors with non-vanishing space-space components, while the medium is stationary, must belong to the field tensor. He then finds that both the Abraham and Minkowski tensors satisfy these requirements. Thus, the values of the electromagnetic momentum flux density are the same in both cases (if the medium is stationary), although the momentum densities are different in the two cases. Because the radiation pressure is determined by the flux density, rather than by the momentum density, the result is the same, regardless of the choice of the (Abraham or Minkowski) tensor.

Mikura, nevertheless, favours the Minkowski expression which, in view of results of experiments [25, 27] should represent the true radiation momentum density. The radiation pressure is then given by the product of this momentum density and the velocity of light in matter, i.e., $(\boldsymbol{D}\times\boldsymbol{B})/4\pi n$, in disagreement with the generally accepted result $(\boldsymbol{E}\times\boldsymbol{H})/4\pi n$. Moreover, Mikura claims that there is no need for the mechanical pressure, $(\boldsymbol{D}\times\boldsymbol{B} - \boldsymbol{E}\times\boldsymbol{H})/4\pi n$, as, in the steady state experiments, the time average effect of the electro- and magneto-strictive forces is balanced by the elastic hydrostatic force. That pressure could be, however, important in transient phenomena.

Mikura's discussion of the acoustic radiation pressure, caused by the density waves, shows it to be smaller than the electromagnetic radiation pressure by a factor $E_0^2/\varrho v_s^2$ (v_s — speed of sound in a medium). As this factor is rather small, the mechanical pressure arising from the density waves is negligible in comparison with the true electromagnetic pressure.

Thus, according to MIKURA [20], although there is no unique way for splitting the total tensor $S^{\mu\nu}$, when the momentum flux density, instead of the momentum density, is taken into account in calculations, the radiation pressure in matter can be uniquely determined.

Noting the fact that the group speed of light waves is reduced in matter and asserting the equivalence of the mass and energy, together with the conservation of the total energy and momentum, THORNBER [74] was led to postulate the existence of transverse forces on matter. Similarly to many workers before him, he was supporting the Abraham hypothesis.

While analyzing the radiation pressure on a multilayer medium, ARNAUD [75] generalized results obtained previously by HAUS [16].

Arguments based on the adiabatic invariance and the energy quantization are used by JOYCE [76] in the analysis of the radiation force. From the operational stand-point, he distinguishes among the impulsive-, the system-, and the energy-carrying momentum. Further, he differentiates between bare and clothed particles and shows that the theory applies not only to photons but also to phonons [77]. In fact, the theory encompasses both zero and non-zero rest mass particles. Thus, he argues that the Abraham momentum refers to a bare particle, whereas that of Minkowski describes the clothed particle. As the essential point of clothing and localization is often confused when interpreting observations, it is, clearly, necessary to distinguish carefully between these concepts. It would appear [76, 77] that, as clothed particles are the only ones observed, the corresponding theory must lead to the Minkowski result.

The classical relativistic field theory served DEWAR [78] as a vehicle to analyze the splitting of the total energy-momentum tensor into its components and also to find the force density acting on a medium in any state. In analogy with the canonical and physical momentum for a particle, Dewar distinguishes between the canonical and physical energy-momentum tensors.

He, first of all, obtains the conservation equations for the whole system using the Noether theorem. This, as is known [79], relates the invariance of equations of motion to the symmetry transformations on the system. It employs two symmetry postulates. The first of these requires the Lagrangian density $\mathscr{L}$ to be invariant under space-time translations. The second postulate requires $\mathscr{L}$ to be invariant under rotation. It is then shown that the angular momentum is conserved and the energy-momentum tensor is symmetrized by the method due to BELINFANTE [47].

The thus obtained symmetrical energy-momentum tensor $\Theta^{\mu\nu}$ is then split up in two alternative ways. The first of these, DEWAR [78] calls the physical split-up. It has the advantage that the background energy-momentum tensor is not modified by the presence of fields, as all interactions are embodied in the term describing the force density on the medium. Hence, the following equations apply to the subsystems

$$\partial_\mu \Theta_b{}^{\mu\nu} = f_b{}^\nu; \qquad \partial_\mu \Theta^{\mu\nu} = f^\nu_{(k)}. \tag{3.73}$$

Here, $f_b{}^\nu$ is the force density acting on the background medium, also known as the ponderomotive force density, while $f^\nu_{(k)}$ is the physical force density acting on the k-th subsystem.

The second way of splitting up the tensor $\Theta^{\mu\nu}$ involves identifying the canonical energy-momentum tensor with the background medium. DEWAR [78] calls this method the modified canonical split-up. Although the tensors corresponding to the constituents of the system are unsymmetrical, the total tensor is symmetric.

Finally, after a long and elaborate investigation, Dewar reaches the conclusion that it is futile to talk about the true electromagnetic momentum, when taking into account the influence of the background medium, as those subsystems are coupled. Consequently, if the canonical energy-momentum tensor is used to describe the background medium, then the Minkowski proposal is correct. On the other hand, when the symmetrical tensor is associated with the medium, then the electro- and magneto-strictive terms must be added to the Abraham tensor, in order to yield correct results.

Not only the electromagnetic momentum in a medium but also the closely related acoustic momentum are rather controversial subjects [76, 77]. Thus, the meaning of the acoustical momentum was discussed by POST [80] utilizing results of ECKART [81]. The question of the energy-momentum tensor was also investigated by JONES [66] who extended the problem to the geomagnetic waves and considered the important concept of the wave spin.

Using a method applicable to the plasma state [82], KLIMA and PETRZILKA [83] proceed to extend the GORDON [12] results. However, while investigating the pressure of the radiation field on a fluid, they do not assume the validity of either the Abraham or the Minkowski expressions, yet arrive at the results consistent with Abraham. Their calculations are, unfortunately, not general, applying only to very dilute gases in which the effects of the order are vanishingly small.

It may be possible, in order to throw some light on the energy-momentum tensor controversy, to regard the asymmetric canonical tensor and the Minkowski tensor as describing the same quantity. Indeed, NOVOBATZKY (cf. [*23*]) has shown that, in the free field, both tensors differ only by the divergence term, so that we may write

$$S_M{}^{\mu\nu} = T^{\mu\nu} + \frac{1}{4\pi}\, \partial_\alpha (A^\mu H^{\nu\alpha}). \tag{3.74}$$

However, this does not hold in the case of an open system. Therefore, it is not possible [*23*], in general, to identify the canonical energy-momentum tensor with a physical tensor supposedly symmetrical. This proposition hinges upon the fact that the equation of conservation of angular momentum applied to a spinless (classical) system leads to the requirement that the energy-momentum tensor be symmetric [*84*].

An interesting way to approach the taks of finding the energy-momentum tensor was suggested by BOPP [*49*]. His investigation is based on the following principles:
1. the field energy flux is proportional to the corresponding momentum; 2. the spin of the radiation field contributes to the total momentum, but not to the energy flux. These requirements were questioned by HEHL [*85*] as they, apparently, did not obey the relativistic invariance. Nevertheless, BOPP [*86*] succesfully defended his original arguments. Thus, the total energy-momentum tensor consists of the linear and the spin parts. The latter then corresponds, in the macroscopic physics, to the phenomenon of magnetization. In other words, the spin-angular momentum, as well as the magnetizing current in macrophysics, are not linked with the energy transport.

3. Experimental results

In the preceding pages the arguments put forward by various workers were discussed purely from the theoretical point of view. It appears that the present situation is unclear, although more workers tend to favour the Abraham tensor. However, the final word in the controversy must be that of the experimentalist. At present, empirical data are rather scarce. Nevertheless, even these do not provide an unambiguous answer to the question of what is the radiation pressure in a medium. Let us now review these experiments.

The most frequently referred to results, relevant to the problem of the radiation momentum in a medium, were obtained by JONES and RICHARDS [*25*]. The experiment consisted, essentially, of the following set up: a mirror, submerged in a refractive liquid, was attached to a torsion balance. Thus, when it was illuminated, incident radiation produced a pressure which, in turn, resulted in a torque on the mirror suspension. Consequently, this torque was measured. Six liquids with different refractive indices were used to investigate the pressure variation as a function of this index. A tungsten lamp served as a radiation source. In order to eliminate the convection currents and the resulting forces (which occur in an unequally heated liquid), Jones and Richards resorted to the use of the radiation chopper. After a careful analysis, these workers claimed that, to within the root mean square error of 1.2%, the radiation pressure is proportional to the refractive index of the liquid. Their conclusion was, therefore, in accord with the Minkowski prediction.

Quite recently, that experiment has been repeated, with an improved accuracy, by JONES and LESLIE [*26*]. In the earlier experiment, due to only about 80% reflectivity of the mirror, the absorbed light was transformed into heat, thus generating convection currents. The differentiation between the forces corresponding to this effect and the radiation forces was, however, possibly due to the unequal time of response in both cases. Now, in order to improve the accuracy, a highly reflecting mirror was used. As a consequence, the minor heating was reduced by the factor of 400. Furthermore, the tungsten lamp of the Jones-Richards experiment was replaced, as the source of illumination, by a He-Ne laser. This served to minimize the uncertainty due to the width of the spectral distribution of the source. The data thus obtained supported the earlier conclusion (as to the validity of the Minkowski expression) with an accuracy of approximately 0.05%.

In addition, it was found that n was the conventional, i.e., the phase- rather than the group-refractive index. Moreover, the Jones-Leslie experiment showed no dependence of the radiation pressure on the plane of polarization for angles of incidence up to 20°. This was contrary to the Peierls prediction (3.63).

At the beginning of this century POYNTING [*87*] suggested that the reflection always produces a normal pressure on a medium, whereas the refraction causes a normal (outward) pull. This was verified experimentally by BARLOW [*88*] when investigating the torque produced by an obliquely incident light beam. The variation of the light torque with the angle between the plane of polarization and plane of incidence, as well as the torque dependence of the wave length were, amongst other things, determined by BETH [*89*]. According to the theory, the torque density $(\boldsymbol{D} \times \boldsymbol{E}/)4\pi$ causes, owing to the high frequency used, a great difficulty in observation. As the wavelength increases, so does the torque. Thus, CARRARA [*90*] used the centimetre electromagnetic waves and obtained a much stronger effect, in accordance with BETH [*89*] findings.

An experiment aimed at determining the transfer of momentum from the radiation field to material particles was performed by RAWSON and MAY [*91*], with additional data collected by MAY, RAWSON, and HARA [*92*]. It was found that the propulsion of dust particles situated in the He-Ne laser beam is due to radiometric forces, due to temperature gradients. Such forces are, usually, orders of magnitude larger than those due to the radiation pressure. It was concluded that, at beam fluxed of 150 W/cm, photophoresis can explain the driving forces and stabilizing torques acting on dust particles.

A similar experiment was performed by ASHKIN [*93*], who investigated the behaviour of micron-size latex spheres in a continuous laser beam. In order to avoid the photophoresis, i.e., the temperature gradients caused by light, particles were suspended in a transparent liquid. Furthermore, the power densities were 10^3 times higher than those reported by RAWSON and MAY [*91*] and MAY et al. [*92*].

It was found, when the spheres were submerged in a liquid with refractive index smaller than that of the spheres, that, in addition to being accelerated along the direction of the beam, they were also drawn toward the center of the laser beam.

Whenever the index of refraction of the spheres was lower than that of surrounding medium (such as the case of air bubbles in the mixture of glycerol and water), the spheres were again accelerated along the direction of the beam but were forced away from the beam axis.

A similar behaviour was postulated to apply also to atoms and molecules in a laser field.

In another experiment by ASHKIN and DZIEDZIC [*27*], the effect of a 4 kW pulsed laser beam on a free surface of a medium was studied. It was found that, after about 390 ns, a positive surface lens developed, thus indicating an outward force around the area of the beam entry into a medium. This also confirmed that, during that time interval, the force acting on a surface must have exceeded the surface tension. Following an additional 700 ns, the lens disappeared, presumably under the delayed influence of surface tension.

Further, when the surface was irradiated from within the liquid, it experienced outward forces. It was concluded, therefore, that both the entrance and exit surfaces experienced forces in the direction away from the medium, thus supporting the Minkowski prediction.

The theoretical explanation of this, as well as of the Jones-Richards experiment, was given by GORDON [12] and later, including the hydrodynamic time response of the liquid surface, by LAI and YOUNG [94]. Also, KATS and KONTOROVICH [95] calculated the influence of laser beam on a liquid surface and, whilst including the non-linear effects, obtained results in agreement with the experiment.

All the above mentioned experiments, measuring the radiation force, employed high (optical) frequencies. They led to the conclusion [25, 26, 27] that correct values were obtained when accepting the Minkowski tensor.

Let us now turn our attention toward the other extreme — the region of very low frequencies and, in the limit, that of a static field.

The electrostatic field interacts with a liquid dielectric and, in turn, causes a compression of the latter. Consequently, the refractive index of the medium increases and can be measured by Schlieren photography. By studying the images thus obtained, the effect of the excess pressure may be observed. This was, essentially, the method used by HAKIM and HIGHAM [30] in their investigation of the force acting on a volume element inside a dielectric. To within an accuracy of 5%, their findings agreed with (3.53). In other words, it was the Abraham tensor that correlated well with the experimental results in the static limit.

This conclusion was also supported by results of JAMES [55]. Using a frequency of about 3 kHz interacting with a ferromagnetic medium, James confirmed that the Minkowski tensor was not applicable at low frequencies.

Still another relevant experiment must be mentioned here. It has not been (to my knowledge) referred to in the literature. That experiment was performed by WALKER et al. [28], using the time varying electric field and a fixed magnetic field. The sample of $BaTiO_3$, a non-magnetic ceramic with a dielectric constant $\varepsilon_r = 3320 \pm 15$, was suspended between the poles of a magnet as a torsional pendulum. The varying electric field with frequency of about 0.3 Hz was then applied to the sample. Whenever that field was exactly in phase with the natural oscillation of the pendulum, the mechanical forces arising from the interaction of the polarization current with the magnetic field would be expected to enhance the oscillations. In the case of phases being opposite, oscillations were damped. If the polarization current was in phase with the varying electric field, the conduction current force cancelled itself over the complete cycle.

The force density predicted from the Abraham tensor is $(\varepsilon\mu - 1)\, \partial_t(\boldsymbol{E} \times \boldsymbol{H})/c^2$, which is extremely small, except at low frequencies. In the case of a constant magnetic field and a non-magnetic sample, it can be expressed as $\partial_t \boldsymbol{P} \times \mu_0 \boldsymbol{H}$. This force, which does not arise in the Minkowski theory, was, in fact, confirmed by experiment [28]. Also, when this experiment was modified [29] so as to include the case of time varying magnetic field, the results obtained were again in agreement with those of the Abraham theory.

In this chapter we have reviewed the main arguments leading to different electromagnetic energy-momentum tensors. Although several distinct forms were proposed, the main discussion centered around the Abraham and Minkowski tensors. Theoretical arguments, as well as the experimental data, were examined and it was found that they did not always correlate. Most authors, nowadays, tend so favour the Abraham tensor, as it is considered to be the more fundamental of the two. Furthermore, in view of the suggestion by BOPP [49] and on realizing that the magnetization (or spin) effects are, essentially, low frequency phenomena, one can say that experiments are, in general, in agreement with the Abraham theory.

In most of the following, the energy-momentum tensor in a plasma medium will be investigated by means of a completely different approach. However, in order to facilitate its understanding, the next chapter will deal with a number of supplementary topics.

IV. Electromagnetic Waves in a Plasma

The discussion in the preceding chapter covered the gaseous, liquid, and solid states of matter, for which the value of the refractive index, at relevant frequencies, was larger than unity. It was seèn, however, that the conclusions reached by different authors varied. Therefore, in an attempt to shed some light on the correct solution, our arguments will revolve around the plasma state. As the dynamics of a real plasma is not always well understood and as it would, furthermore, unnecessarily complicate the present investigation, our main efforts will be concerned with an approximation, namely, an isotropic, homogeneous, collisionless plasma. The aim of this chapter will be, therefore, to set out the general properties of the fourth state of matter (in that approximation).

The concept of a plasma refers to a conglomeration of charged and neutral particles, there being approximately equal number densities of positive and negative charges. Often, at least one of these charges may be regarded as highly mobile. However, in order to reproduce typical plasma characteristics, it is necessary for the average distances between the charged plasma particles to be very much smaller than the actual physical dimensions of the plasma region. Thus, for instance, a metallic solid, with its positive charges fixed and the concentration of the conduction electrons reading values of the order of 10^{23} cm^{-3}, can often display plasma characteristics. Similarly, one can conceive of a liquid plasma. In the present work, however, our investigation will be directed mainly towards tenuous, gaseous plasmas.

From the macroscopic point of view, that state is characterized by several concepts. Thus, if subscripts e and i refer to electrons and ions, respectively, and n, q, m, in turn, correspond to the average number density, charge, and mass of these charge carriers, then it follows that

$$\bar{n}_e q_e + \bar{n}_i q_i = 0,$$

provided that the plasma is taken to be electrically neutral. Moreover, the average thermal speeds of the plasma constituents are zero, i.e., $\bar{v}_e = \bar{v}_i = 0$. Now, in analogy to the kinetic theory of gases, the motion of these particles results in a pressure. That (kinetic) pressure can be expressed by

$$\bar{p}_e = \bar{n}_e \varkappa T, \qquad \bar{p}_i = \bar{n}_i \varkappa T. \tag{4.1}$$

Here, $\varkappa$ is the Boltzmann constant and T denotes the temperature. In the case of thermodynamic equilibrium, there applies the equality $T_e = T_i$. However, the temperatures T_e and T_i become distinct when external fields are applied to the plasma. If the external magnetic field B_0 is transverse to the particle motion, then the charged particles execute a circular motion. The corresponding (so called cyclotron) angular frequencies for electrons and ions are then, respectively,

$$\omega_b = -q_e B_0/m_e; \qquad \Omega_b = q_i B_0/m_i. \tag{4.2}$$

If, on the other hand, the velocity of the particle is parallel to the field, it cannot be affected by that field. In general, therefore, the particles travel along a helical trajectory.

Another characteristic of plasma state is its tendency to preserve charge neutrality. Hence, microscopic charge imbalances, due to thermal motion, are opposed by the induced electric field. Thus, electrons oscillate in the direction of the electric field vector, which, in turn, is parallel to the direction of propagation. This kind of plasma electron oscillations is known as the longitudinal oscillation. Parameters associated with this behaviour are called the electron (ω_p) and the ion (Ω_p) plasma frequencies and these are defined by

$$\omega_p^2 = 4\pi\bar{n}_e q_e^2/m_e; \qquad \Omega_p^2 = 4\pi\bar{n}_i q_i^2/m_i. \tag{4.3}$$

The plasma state is capable of supporting a number of thermally and externally induced oscillations. For instance, low frequency electromagnetic waves in the presence of an external magnetic field, also known as the hydromagnetic waves, can propagate in a plasma with speeds often many times smaller than that of light in vacuo. Further, not only the transverse but also longitudinal waves occur in plasmas, as will be seen below.

In most cases, the wavelength λ of the electromagnetic radiation is much greater than the interparticle distance in plasma. Therefore, it is possible to analyze the wave propagation in a plasma from the point of view of the phenomenological electrodynamics. In so far as the concepts of charge density ϱ and current density $\boldsymbol{j}$ are involved, the plasma is considered as an equivalent to a continuum. In other words, the discreteness of the plasma constituents is ignored and, hence, the Maxwell equations are used to describe smeared-out, average fields and not the microscopic, interparticle fields. If these macroscopic, electric and magnetic field vectors are $\boldsymbol{E}$ and $\boldsymbol{H}$, with $\boldsymbol{D}$ and $\boldsymbol{B}$ denoting the corresponding inductions, then the Maxwell field equations yield

$$\begin{aligned} \nabla \times \boldsymbol{E} &= -\frac{1}{c}\,\partial_t \boldsymbol{H}, \qquad \nabla \times \boldsymbol{H} = \frac{4\pi}{c}\,\boldsymbol{j} + \frac{1}{c}\,\partial_t \boldsymbol{D}, \\ \nabla \cdot \boldsymbol{H} &= 0, \qquad \nabla \cdot \boldsymbol{D} = 4\pi\varrho. \end{aligned} \tag{4.4}$$

Note that, as the magnetic permeability of a tenuous plasma is practically equal to unity, the magnetic induction $\boldsymbol{B}$ was replaced, in (4.4), by the magnetic field $\boldsymbol{H}$. Now, to make use of (4.4), we need the actual $\boldsymbol{E}$-dependence of the electric induction $\boldsymbol{D}$ and current density $\boldsymbol{j}$. In absence of any external magnetic field and for an isotropic plasma, we have

$$\boldsymbol{D} = \varepsilon \boldsymbol{E}, \qquad \boldsymbol{j} = \sigma \boldsymbol{E}, \tag{4.5}$$

where ε and σ are, respectively, the dielectric permittivity and the conductivity of the plasma. The presence of an external magnetic field disturbs the plasma isotropy, so that (4.5) is modified to

$$D_k = \varepsilon_{kl} E_l \qquad j_k = \sigma_{kl} E_l, \tag{4.5a}$$

where ε_{kl} and σ_{kl} are now tensor quantities corresponding to ε and σ. If the magnetic field is assumed to be time independent, relation (4.5a) can be regarded as linear. Fortunately, this approximation is reasonably valid in a majority of cases. However, in very strong fields, (4.5a) becomes non-linear, so that, as a result, interactions appear among the various kinds of waves.

It is convenient to combine the ε_{kl} and σ_{kl} tensors into a single, complex permittivity tensor

$$\tilde{\varepsilon}_{kl} = \varepsilon_{kl} - 4\pi i \sigma_{kl}/\omega, \tag{4.6}$$

which, in general, depends on all coordinates.

For the sake of simplicity, consider now a homogeneous plasma, i.e., a plasma in a region where $\tilde{\varepsilon}_{kl}$ is constant in space. Further, assume that a plane monochromatic wave propagates in the positive z-direction and at normal incidence to the plasma boundary. Whilst deriving the explicit form of the dispersion relation, it will be seen that, in the present approximation, the plasma is able to support four different types of oscillations. A very clear exposition pertinent to this topic was given by DENISSE and DELCROIX [96] and we shall follow it rather closely here.

Let us first eliminate the magnetic field strength from (4.4). Assuming that all variables are simple harmonic functions of time (i.e., are proportional to $e^{i\omega t}$) we get

$$\nabla \times \nabla \times \boldsymbol{E} = \frac{4\pi i}{c^2}\,\omega \boldsymbol{j} + \frac{\omega}{c^2}\,\boldsymbol{D}. \tag{4.7}$$

But, the current density $\boldsymbol{j}$ can be written as

$$\boldsymbol{j} = \bar{n}_e q_e v_e + \bar{n}_i q_i v_i\,. \tag{4.8}$$

Hence, on substituting the plane monochromatic wave in (4.7) and utilizing (4.5) and (4.8), we obtain

$$\begin{aligned} \big((k^2c^2/\omega^2) - 1\big)\, E_x &= -4\pi i j_x/\omega \\ \big((k^2c^2/\omega^2) - 1\big)\, E_y &= -4\pi i j_y/\omega \\ -E_z &= -4\pi i j_z/\omega\,, \end{aligned} \tag{4.9}$$

where $i = \sqrt{-1}$. Thus, it is clear from (4.9) that E_z is always lagging with respect to j_z, whilst E_x and E_y can lag or advance relative to the respective current components, depending on whether $(k^2c^2/\omega^2) - 1$ is negative or positive.

Next, the conservation equations are determined. It will be assumed that the perturbation of the plasma variables caused by the passage of the wave, is small and can be neglected.

If the symbols n_e and n_i denote the fluctuations of the number density of electrons and ions, respectively, then the corresponding conservation equations are written as

$$\partial_t n_e + \nabla \cdot \bar{n}_e \boldsymbol{v}_e = 0; \qquad \partial_t \bar{n}_i + \nabla \cdot \bar{n}_i \boldsymbol{v}_i = 0\,. \tag{4.10}$$

The momentum conservation equations are then, on neglecting the effect of the magnetic field of the wave on particles,

$$\begin{aligned} \bar{n}_e m_e(\partial_t \boldsymbol{v}_e) &= \bar{n}_e q_e(\boldsymbol{E} + \boldsymbol{v}_e \times \boldsymbol{B}_0) - \nabla p_e + \boldsymbol{P}_{en} + \boldsymbol{P}_{ei} \\ \bar{n}_i m_i(\partial_t \boldsymbol{v}_i) &= \bar{n}_i q_i(\boldsymbol{E} + \boldsymbol{v}_i \times \boldsymbol{B}_0) - \nabla p_i + \boldsymbol{P}_{in} + \boldsymbol{P}_{ei}\,. \end{aligned} \tag{4.11}$$

We assumed here that the phase speed of the wave is much greater than the thermal speed; for otherwise, these momentum equations would be incomplete. The terms $\boldsymbol{P}_{xy}$, with subscripts e, n, i, and their permutations, refer to momentum transfers in collisions involving electrons, neutral particles, and ions. It is assumed [96] that they can be expressed as

$$\begin{aligned} \boldsymbol{P}_{en} = -\bar{n}_e m_e \nu_{en} v_e; \qquad \boldsymbol{P}_{in} &= -\bar{n}_i m_i \nu_{in} \boldsymbol{v}_i; \qquad \boldsymbol{P}_{ei} = -\bar{n}_e m_e \nu_{ei}(\boldsymbol{v}_e - \boldsymbol{v}_i) \\ \boldsymbol{P}_{ei} &= -\boldsymbol{P}_{ie}\,. \end{aligned} \tag{4.12}$$

The symbol ν, with appropriate subscripts, refers to the collision frequency between the corresponding plasma constituents. Although this frequency is assumed here to be a constant, in general it also has a tensor character. Further, following [96], the pressure gradient is written in terms of the average thermal speeds V_γ [97], where $V_\gamma{}^2 = 3\varkappa T_\gamma/m_\gamma$, as

$$\nabla p_\gamma = m_\gamma V_\gamma \nabla n_\gamma\,,$$

with subscript γ referring to either electrons or ions. Asserting [96] that the particle conservation may be written as

$$n_\gamma = k\bar{n}_\gamma v_{\gamma z}/\omega$$

which, after substitution in the above, yields

$$\nabla p_\gamma = -ik\bar{n}_\gamma m_\gamma V_\gamma{}^2 v_{\gamma z} \boldsymbol{k}/\omega . \tag{4.13}$$

With the assumption that the magnetic field can be split into two components, one, B_T, normal and the other, B_L, parallel to the direction of propagation, the momentum conservation equations can be written out more explicitly. Thus, the substitution of (4.13) and (4.12) into (4.11) leads to conservation equations in the component form

$$\begin{aligned} i\omega\bar{n}_\gamma m_\gamma v_{\gamma x} &= \bar{n}_\gamma q_\gamma(E_x + v_{\gamma y}B_L - v_{\gamma z}B_T) \mp \nu_{ei}\bar{n}_\gamma m_\gamma(v_{ex} - v_{ix}) - \nu_{\gamma n}\bar{n}_\gamma m_\gamma v_{\gamma x} \\ i\omega\bar{n}_\gamma m_\gamma v_{\gamma y} &= \bar{n}_\gamma q_\gamma(E_y - v_{\gamma x}B_L) \mp \nu_{ei}\bar{n}_\gamma m_\gamma(v_{ey} - v_{iy}) - \nu_{\gamma n}\bar{n}_\gamma m_\gamma v_{\gamma y} \\ i\omega\bar{n}_\gamma m_\gamma v_{\gamma z} &= \bar{n}_\gamma q_\gamma(E_z + v_{\gamma x}B_T) \mp \nu_{ei}\bar{n}_\gamma m_\gamma(v_{ez} - v_{iz}) - \nu_{\gamma n}\bar{n}_\gamma m_\gamma v_{\gamma z} + ik^2\bar{n}_\gamma m_\gamma V_\gamma{}^2 v_{\gamma z}/\omega . \end{aligned} \tag{4.14}$$

When the above equations describe electron (ion) momentum conservation, then the negative (positive) sign before the electron-ion collision term is used. In conjugation with (4.9), they form a set of nine homogeneous linear equations with nine unknown. These unknowns are components of vectors $\boldsymbol{E}$, $\boldsymbol{v}_i$, and $\boldsymbol{v}_e$. Hence, in order that such a set has a solution, the determinant formed from the coefficients of the individual terms must be equal to zero. If this is the case, there results an equation of the fourth order in k. It, therefore, means that the system has four solutions which describe four, generally different, modes of propagation: the ordinary, extraordinary, electron, and ion modes [*98*]. Moreover, there are two conjugate solutions for each of k^2, which means that the (identical) waves propagate in the opposite directions. When setting up this determinant, the collision frequency is, for simplicity, put equal to zero. Then, the first column refers to the variable E_x, second column — to v_{ex}, third column — to v_{ix}, and so on, for the other two components. In addition the first equation of (4.9) and (4.14), referring to electrons and ions, occupy the first, second, and third rows, respectively. The y- and z-components follow in the same order — up to row 9. This determinant is written out in [*96*]. Using the rules applicable to determinants, a more manageable form describing the general dispersion relation is obtained

$$\begin{vmatrix} 1-\frac{k^2c^2}{\omega^2} & \frac{\omega_p{}^2}{\omega^2} & \frac{\Omega^2}{\omega^2} & 0 & 0 & 0 & 0 & 0 & 0 \\ 1 & 1 & 0 & 0 & -\frac{\omega_L}{\omega} & 0 & 0 & \frac{\omega_T}{\omega} & 0 \\ 1 & 0 & 1 & 0 & 0 & \frac{\Omega_L}{\omega} & 0 & 0 & -\frac{\Omega_T}{\omega} \\ 0 & 0 & 0 & 1-\frac{k^2c^2}{\omega^2} & \frac{\omega_p{}^2}{\omega^2} & \frac{\Omega_p{}^2}{\omega^2} & 0 & 0 & 0 \\ 0 & -\frac{\omega_L}{\omega} & 0 & 1 & 1 & 0 & 0 & 0 & 0 \\ 0 & 0 & \frac{\Omega_L}{\omega} & 1 & 0 & 1 & 0 & 0 & 0 \\ 0 & 0 & 0 & 0 & 0 & 0 & 1 & \frac{\omega_p{}^2}{\omega^2} & \frac{\Omega_p{}^2}{\omega^2} \\ 0 & 0 & \frac{\omega_T}{\omega} & 0 & 0 & 0 & 1 & 1-\frac{k^2V_e{}^2}{\omega^2} & 0 \\ 0 & 0 & -\frac{\Omega_T}{\omega} & 0 & 0 & 0 & 1 & 0 & 1-\frac{k^2V_i{}^2}{\omega^2} \end{vmatrix} = 0 . \tag{4.15}$$

In the dispersion relation (4.15), symbols $\omega_T(\Omega_T)$ and $\omega_L(\Omega_L)$ refer to the electron (ion) cyclotron frequency resulting when the magnetic field is split into its transverse and longitudinal components. The relation (4.15), although appearing to be rather complicated, readily lends itself to simplification.

Consider first a plasma in absence of any external magnetic field, $H_0 = 0$. In this case, both electron and ion cyclotron frequencies vanish. The dispersion relation then reduces to

$$A_T{}^2 A_L = 0, \tag{4.16}$$

where A_T, A_L are 3×3 subdeterminants of (4.15), i.e.,

$$\begin{vmatrix} A_T & 0 & 0 \\ 0 & A_T & 0 \\ 0 & 0 & A_L \end{vmatrix} = 0.$$

In view of its derivation, (4.16) expresses the relation between ω and k, i.e., the dispersion relation, in the absence of the external field. Thus, to find a non-trivial solution of (4.16), we have either $A_T{}^2 = 0$ or $A_L = 0$. Consider the former case first. It consists of a product of two identical determinants of order three. Due to the construction of (4.16), the first determinant, $A_T{}^{(1)}$, corresponds to the x-component, whilst the second one, $A_T{}^{(2)}$, describes the y-component. To emphasize again, the direction of propagation is along the z-axis. If the first determinant is now set equal to zero, all components other than E_x, v_{ex}, v_{ix} vanish. This, therefore, means that the thus obtained result informs us about the x-component of the field and of the particle velocity. Put otherwise, if this determinant vanishes, we say that the wave is polarized in the x-direction.

Similarly, when the second determinant, $A_T{}^{(2)}$, is set equal to zero, the obtained solution for E_y, v_{ey}, v_{iy} immediately implies that the wave is polarized in the y-direction.

It is clear, therefore, that the equation $A_T{}^2 = 0$ determines the transverse characteristics of the electromagnetic wave.

Concentrate now on the second equation implied in (4.16), i.e., $A_L = 0$. In that case, the only non-zero roots are E_z, v_{ez}, and v_{iz}. This then means that the wave is polarized in the direction of propagation and, hence, it is a longitudinal wave. Because the equation corresponding to the developed determinant is quadratic in k^2, there are two roots, thus implying two types of longitudinal wave.

The above conclusions will be mainly used in the following chapters of the present work. However, to help us form on overall picture of the wave propagation in plasma and to deepen our insight into the restrictive assumptions used, other types of waves are introduced in the next few paragraphs.

The presence of a magnetic field parallel to the direction of the wave propagation, does not affect the longitudinal waves, as a glance at (4.15) will reveal. Both the longitudinal electron and ion cyclotron frequencies have now non-zero values, ensuring a coupling between the above mentioned waves which then leads to two distinct modes of propagation. The plasma behaves like an anisotropic, doubly refracting medium and the two waves travel with different velocities. This lends itself to a simple physical interpretation: the plane electromagnetic wave can be considered to consist of two (right- and left-) circularly polarized waves. The properties of those separate waves are determined from their interactions with plasma particles. The electric field of an electromagnetic wave alters the orbiting trajectories of electrons and ions (induced by the static magnetic field) in a way dependent on the relation between the wave and the cyclotron frequencies. Thus, the left circularly polarized wave rotates in the same direction as would the ions under the influence of the static external magnetic field. The other polarization agrees with the direction of motion of an electron in the magnetic field. Hence, the electro-

magnetic wave, upon entering the plasma, encounters different particle behaviour, thus leading to an analogy with a doubly refracting crystal.
To generalize this investigation still further, the external magnetic field will be next assumed to be in an arbitrary direction. The resulting dispersion relation becomes, accordingly, very involved. In this case, the transverse waves are coupled to the longitudinal ones, leading to a rather complicated behaviour [*96*]. This interaction, in turn, results in four distinct modes of propagation and one can no longer talk either about a pure transverse or a pure longitudinal wave. Thus, only mixed modes exist and their discussion is not immediately necessary for our present task.
It follows from the preceding discussion that, in the case when the transverse magnetic field is zero, there are two transverse and two longitudinal modes. Utilizing that part of the determinant which corresponds to the transverse waves implies that these modes of propagation do not depend on the thermal motion of particles.
Consider now a plane monochromatic wave normally incident on a plasma. A simple manipulation of the Maxwell equations then leads to

$$\Delta \boldsymbol{E} + \left(\omega^2 \tilde{\varepsilon}(\omega) \boldsymbol{E}/c^2\right) = 0. \tag{4.17}$$

As the electromagnetic field is transverse, the above equation applies to both E_x and E_y components. The third component can be found from

$$\tilde{\varepsilon}(\omega) E_z = 0. \tag{4.18}$$

If $\tilde{\varepsilon}(\omega) \neq 0$, then $E_z = 0$ satisfies (4.18) and the obtained answer is associated with the pure transverse waves. On the other hand, if $\tilde{\varepsilon}(\omega) = 0$ and $E_z \neq 0$, one refers to a longitudinal wave in an isotropic, homogeneous plasma. The root of $\tilde{\varepsilon}(\omega) = 0$ leads, generally, to a complex frequency, meaning that the wave is evanescent. The associated damping is in most cases only slight, since the imaginary part of $\tilde{\varepsilon}(\omega)$ is small relative to the real part. This approximation is valid for a simple plasma, so that $\tilde{\varepsilon}(\omega) = 0$ may then be replaced by

$$\varepsilon(\omega) = 0. \tag{4.19}$$

The solution of (4.19) has a real root, the so called plasma frequency, mentioned above. If the plasma oscillations are simple harmonic, with the phase factor $\exp\left(i(\omega t - \boldsymbol{k} \cdot \boldsymbol{r})\right)$, the frequency ω and the wave vector $\boldsymbol{k}$ are unrelated. This, clearly, is so, as (4.19) determines only the frequency. Therefore, the group speed of these waves is zero, thus implying no energy transport. This presumed behaviour (which does not agree with the experimental data) results from our neglect of the spatial dispersion. The ratio of the Debye length and the wavelength in the medium determines this effect and its small magnitude permits, often, to omit it.
In the approximation used here, the longitudinal waves are unconnected with the transverse waves [*96*]. Hence, two types of waves can be regarded as independent, except in a region where the wave frequency has a value close to that of plasma frequency. The separation of the field into transverse and longitudinal modes cannot, however, be effected in an anisotropic plasma.
As implied above, the permittivity is a rather important concept, as it establishes the link between the electric induction and the electric field. It has been, generally, assumed [*31*], that either the effective field $\boldsymbol{E}_{\text{eff}}$ is equal to the external field $\boldsymbol{E}$, or it is modified by the presence of the Lorentz polarization term, $4\pi\boldsymbol{P}/3$. The differentiation, however, between those two possibilities proves to be not a trivial matter. The question has attracted a lot of attention and there exists diversity of opinions [*31, 32, 33, 99, 101*].

The overall charge of plasma vanishes. It means that the charge of the free electrons must somehow be neutralized. Two possible ways in which this may happen have been put forward: by a positive continuum and by the discrete positive ions. Difficulties arise in attempting to determine the conditions under which these two approaches may be equivalent.

A detailed analysis [*31*] reveals that $\boldsymbol{E}_{\text{eff}} = \boldsymbol{E}$ inside a medium in which electrons are neutralized by a positive continuum. On the other hand, a medium consisting of a set of bound atoms obeys the Lorentz formula

$$\boldsymbol{E}_{\text{eff}} = \boldsymbol{E} + \frac{4\pi}{3}\boldsymbol{P}.$$

It, therefore, follows that, when considering a fully ionized plasma, the correction term $4\pi\boldsymbol{P}/3$ may be neglected [*32*, *33*].

The main contribution, in determining the value of dielectric permittivity, arises from the motion of charges. The effect of neutral particles is important only when the ionization is very small. In the case under consideration, it may be, therefore, neglected. In order to obtain accurate expression for ε and σ, one must start with the Boltzmann equation. However, as most of the present work applies to the case of cold plasma, an approximate derivation of the permittivity will be sufficient.

The high conductivity of plasma is well known. It signifies that, although macroscopically neutral, a plasma consists of moving free charged particles. The total current density $\boldsymbol{j}_t$, due to this motion, is then written as

$$\boldsymbol{j}_t = q \sum_{k=1}^{N} \dot{\boldsymbol{r}}_k - \dot{\boldsymbol{r}}_k^{(i)}, \tag{4.20}$$

where q is the electronic charge, $\dot{\boldsymbol{r}}_k$ and $\dot{\boldsymbol{r}}_k^{(i)}$ are the time derivatives of the electron and ion position vectors, respectively.

Now, the equation of motion of the electron in a collisionless, isotropic plasma is

$$m\ddot{\boldsymbol{r}}_k = qE_0 e^{i\omega t} = q\boldsymbol{E}, \tag{4.21}$$

where $\boldsymbol{E}$ denotes the mean macroscopic field, in agreement with the above discussion. The solution of (4.21) is

$$\boldsymbol{r}_k = -(q\boldsymbol{E}/m\omega^2) + \boldsymbol{r}_k'(t), \tag{4.22}$$

where $\boldsymbol{r}_k'(t)$ is the position vector in the absence of the field. When (4.21) describes the ion motion, then the corresponding solution is

$$\boldsymbol{r}_k^{(i)} = -(q\boldsymbol{E}/M\omega^2) + \boldsymbol{r}_k^{(i)\prime}(t) \tag{4.23}$$

and M stands for the ion mass.

Now, by definition, the total current density is

$$\boldsymbol{j}_t = \boldsymbol{j} + i\omega\boldsymbol{P} = \left(\sigma + i\frac{\varepsilon - 1}{4\pi}\,\omega\right)\boldsymbol{E}. \tag{4.24}$$

Substitution of (4.22) and (4.23) into (4.24) shows then that the conduction vanishes and the polarization vector can be written as

$$\boldsymbol{P} = -\frac{q^2\boldsymbol{E}}{\omega^2}\left(\frac{n_e}{m} + \frac{n_i}{M}\right).$$

As there is no polarization in the absence of the field, then $\sum r_k - r_k^{(i)} = 0$ and, hence, the permittivity can be expressed as

$$\varepsilon = 1 - \frac{4\pi q^2}{\omega^2}\left(\frac{n_e}{m} + \frac{n_i}{M}\right),$$

where n_e and n_i refer to electron and ion densities, respectively. On account of their mass, ions usually have only small influence of the permittivity and are, therefore, often neglected. Thus, the approximate expression for the permittivity becomes

$$\varepsilon = 1 - (\omega_p{}^2/\omega^2), \tag{4.25}$$

where $\omega_p{}^2 = 4\pi q^2 n_e/m$ defines the plasma frequency.

As already indicated the conductivity is zero under the above assumptions. This is so, because here we refer to a collisionless plasma. Hence, electrons do not transfer energy but merely oscillate in the field. These conditions change when collisions are included. Then, a term proportional to the frictional force, $\alpha\dot{\boldsymbol{r}}$, is added to the right hand side of (4.21). This force is caused by the momentum transfer occurring during collisions. On repeating, essentially, the same arguments as above, we obtain the result

$$\tilde{\varepsilon} = 1 - \frac{\omega_p{}^2}{\omega(\omega - i\nu_{\text{eff}})}, \tag{4.26}$$

where ν_{eff} is the effective collision frequency.

The complex permittivity implies the existence of a non-vanishing conductivity and, hence, the absorption of energy. It must be stressed, however, that (4.26) is only an approximation, and the exact result, obtained from the Boltzmann theory, cannot be reduced [*33*] to it. Nevertheless, the relation (4.26) is quite adequate for our purposes.

The question now arises as to the validity of applying the classical theory of mechanics to the motion of electrons and ions. It is found [*33*], that the classical theory is valid if $\hbar\omega \ll mc^2$, where $2\pi\hbar$ is the Planck constant. Another condition for the applicability of the classical theory is that the plasma be regarded only in a non-degenerate case.

When analyzing the propagation of the electromagnetic waves in plasma (or, generally, in a medium) the concepts of the refractive and absorption indices become more prominent. Let us elaborate this point and find out the influence of these indices on some variables pertinent to propagation. Thus, the electromagnetic wave equation in a medium with a constant permittivity tensor can be written [*8*] as

$$\Delta\boldsymbol{E} + (\omega^2/c^2)\,\tilde{\varepsilon}(\omega)\,\boldsymbol{E} = 0. \tag{4.27}$$

If a plane electromagnetic wave, $\boldsymbol{E} = E_0 \exp\big(i(\omega t - \boldsymbol{k}\cdot\boldsymbol{r})\big)$ is substituted into (4.27), one obtains the dispersion relation

$$k^2 = (\omega^2/c^2)\,\tilde{\varepsilon}(\omega). \tag{4.28}$$

To obtain the relation between $\boldsymbol{E}$ and $\boldsymbol{H}$, the plane wave expression is inserted into the two Maxwell equations (4.4) which involve the curl operator. This yields

$$\omega\tilde{\varepsilon}(\omega)\,\boldsymbol{E} = -c\boldsymbol{k}\times\boldsymbol{H} \qquad \omega\boldsymbol{H} = c\boldsymbol{k}\times\boldsymbol{E}. \tag{4.29}$$

In order to find the condition for transverse waves, (4.29) is (scalarly) multiplied by $\boldsymbol{k}$, giving $\boldsymbol{k}\cdot\boldsymbol{E} = 0$ and $\boldsymbol{k}\cdot\boldsymbol{H} = 0$. It is usually assumed that $\tilde{\varepsilon}(\omega) \neq 0$, otherwise, the longitudinal waves may appear, as indicated above. When the wave vector is complex, whilst the frequency is real, then the planes of equal phase differ from the planes of equal amplitude, resulting in inhomogeneous waves. On the other hand, when the waves are homogeneous, those two planes coincide. The vector may then be written as [8]

$$\boldsymbol{k} = (\omega/c)\,(n - i\alpha)\,(\boldsymbol{k}/k), \tag{4.30}$$

where n and α are the indices of refraction and absorption, respectively. Replacing the $\boldsymbol{k}$-vector in the plane wave expression by (4.30) leads to

$$\boldsymbol{E} = \boldsymbol{E}_0 e^{-\omega\alpha z c} e^{i(\omega t - \omega n z/c)}. \tag{4.31}$$

Thus, the simple harmonic wave is modulated by the factor $\exp(-\omega\alpha z/c)$, which is connected with the absorption process. As (4.31) implies, the amplitude of the wave in a medium changes by a factor of e over the distance $\lambda_0/2\pi\alpha$ (λ_0 the vacuum wavelength). Now, utilizing (4.31), the relation between the wavelength λ in a medium to that λ_0 in vacuo becomes

$$\lambda = \frac{\lambda_0}{n}. \tag{4.32}$$

In the absence of absorption, when only the scattering is present, the refractive index n is equal to $\sqrt{\varepsilon}$. This no longer holds when there is also some absorption. Thus, comparing (4.30) with (4.6) leads to

$$\tilde{\varepsilon} = (n - i\alpha)^2$$

or

$$\varepsilon = n^2 - \alpha^2 \qquad 2n\alpha = \frac{4\pi\sigma}{\omega}. \tag{4.33}$$

Whence,

$$n = \left[\frac{1}{2}\,\varepsilon + \left(\left(\frac{1}{2}\,\varepsilon\right)^2 + \left(\frac{2\alpha\sigma}{\omega}\right)^2\right)^{1/2}\right]^{1/2} \tag{4.34}$$

and

$$\alpha = \left[-\frac{\varepsilon}{2} + \left(\left(\frac{\varepsilon}{2}\right)^2 + \left(\frac{2\pi\sigma}{\omega}\right)^2\right)^{1/2}\right]^{1/2}.$$

Applying expressions (4.34) to a collisionless plasma with parameters $\varepsilon = 1 - (\omega_p^2/\omega^2)$ and $\sigma = 0$ yields

$$\begin{aligned} n &= \left(1 - \frac{\omega_p^2}{\omega^2}\right)^{1/2}, & \alpha &= 0 & &\text{if} \quad \omega > \omega_p \\ n &= 0, & \alpha &= \left(\frac{\omega_p^2}{\omega^2} - 1\right)^{1/2} & &\text{if} \quad \omega < \omega_p \\ n &= 0, & \alpha &= 0 & &\text{if} \quad \omega = \omega_p. \end{aligned} \tag{4.35}$$

Thus, the characteristics of wave propagation differ dramatically, depending on whether the angular frequency ω of the wave is larger or smaller than the plasma frequency. Hence, when $\omega > \omega_p$, the wave travels without attenuation, whilst in the case $\omega < \omega_p$, the wave in a medium becomes evanescent, i.e., non-absorptively damped. Put otherwise, the wave cannot propagate in a medium and is completely reflected when $\omega \leqq \omega_p$. Then, according to the Mawxell equations $\nabla \times \boldsymbol{E} = 0$ and the magnetic field vanishes. As a consequence, the transverse electromagnetic waves cannot exist. However, under these conditions, the longitudinal waves may appear.

The wave vector $\boldsymbol{k}$ whose direction indicates the direction of propagation, is related to the wavelength by $k = 2\pi/\lambda$. Therefore, taking into account (4.32), the wave vector in a medium is related to that in vacuo by

$$\boldsymbol{k} = n\boldsymbol{k}_0. \tag{4.36}$$

There is a close connection between (4.36) and the concept of the phase and group velocities. Thus, the phase velocity is defined as describing the propagation of phase. In other words, it refers to the phase differences between oscillations observed at two different points. Hence, when the frequency ω and the wave vector $\boldsymbol{k}$ are given, the phase speed is expressed as

$$v_p = \frac{\omega}{k}. \tag{4.37}$$

This speed is primarily [*101*] used in the treatment of standing waves and in the interference and diffraction phenomena, as phase differences are, then, of particular importance.
The signal, or modulation of the wave, on the other hand, travels (when absorption is small) with the group velocity. The group speed, v_g, is, in a dispersive medium, different from v_p and is defined by

$$v_g = \frac{d\omega}{dk}. \tag{4.38}$$

Hence, in vacuum, which is, of course, non-dispersive to electromagnetic waves, $v_p = v_g = c$ and the signal propagates without distortion. To emphasize again, the phase velocity has nothing to do with propagation [*101*], but gives us, essentially, the distribution of phase in space. On the other hand, the group velocity refers, in the present approximation, to the energy transport. However, when absorption is high, the signal is strongly frequency dependent, becomes distorted, and the concept of the group velocity then loses its meaning [*101*].
Let us now apply these concepts to an isotropic, homogeneous plasma. Consider an electromagnetic wave of frequency $\omega > \omega_p$. Then, assuming absence of absorption, it follows from (4.30) that the wave vector is real. Its subsequent substitution in (4.37) and (4.38) then determines the phase and group speeds of waves, respectively. Thus

$$v_p = \frac{c}{n} \qquad v_g = cn, \tag{4.39}$$

which means, as $n < 1$ in plasma, that $v_p > c$ and $v_g < c$. However, as the phase velocity is not associated with the energy transport, it does not contradict the theory of relativity. At frequencies much larger than the plasma frequency, both speeds approach the vacuum value of light, as is obvious from (4.35).
As already indicated, when $\omega < \omega_p$, the value of the wave vector is imaginary, the wave exponentially decays, penetrates to a certain thickness, known as the skin depth, and is then totally reflected.
The above described behaviour can be explained in relatively simple terms. At very low frequencies, electrons in a plasma have enough time to move in such a way as to screen the plasma interior from the oscillating external electric field. In other words, it follows from the Maxwell equations that the real current density, caused by the motion of charged particles, is greater than the displacement current density. At very high frequencies, on the other hand, electrons cannot move fast enough to follow the oscillating field and, thus, are unable to screen the plasma interior. In other words, the real current

density is small compared with the displacement current density, due to the inertia of the electrons. Therefore, the behaviour of an electromagnetic wave of high frequency in plasma approaches the one in vacuo.
It is of interest to inquire how the propagation is affected when there is a small absorption. Thus, in the limit $|\bar{\varepsilon}| \gg 4\pi\sigma/\omega$, relations (4.34) reduce to

$$n \approx \left[1 - \frac{\omega^2{}_p}{(\omega^2 + \nu_{\text{eff}}^2)}\right]^{1/2}$$
$$\alpha \approx \frac{\omega_p{}^2\nu_{\text{eff}}}{\left[2\omega(\omega^2\nu_{\text{eff}}^2)\left(1 - \left(\frac{\omega_p{}^2}{\omega^2 + \nu_{\text{eff}}^2}\right)\right)\right]^{1/2}}. \quad (4.40)$$

The corresponding phase and group speeds are then, respectively,

$$v_p = \frac{c}{\left(1 - \frac{\omega_p{}^2}{\omega^2 + \nu_{\text{eff}}^2}\right)^{1/2}}$$
$$v_g = \frac{c[1 - \omega_p{}^2/(\omega^2 + \nu_{\text{eff}}^2)]^{1/2}}{1 - \omega_p{}^2\nu_{\text{eff}}^2/(\omega^2 + \nu_{\text{eff}}^2)^2}. \quad (4.41)$$

Comparing (4.41) with (4.39) confirms that, in the presence of absorption, the phase speed is decreased, whereas the group speed increases.
Considerations in this chapter should serve to provide sufficient information to form a background for analysis in following chapters. Some of the features presented here will be further developed there.

V. Dual Nature of Radiation

The complementary character of the corpuscular and undulatory aspects of electromagnetic radiation is, nowadays, readily accepted. Only the two aspects taken together can reveal the complete nature of that radiation.
Very often, in treating radiation phenomena, concepts of the photon and the wave packet (meant to represent the two complementary kinds of radiation) are used interchangeably. This is, of course, quite permissible in the empty space. It is believed, however, that such a (random) alternation between these two notions cannot be validly applied when the radiation propagates in a material medium. Thus, Skobel'tsyn [*35*] refers to a "quantum photon" and a "classical photon", in order to discriminate between the inherent differences of the two aspects. It is a common mistake to confuse the corpuscular and undulatory characters of radiation, when it is inside a medium. It should be, therefore, emphasized that these two notions represent two distinct (but complementary) aspects of the interaction phenomenon.
Let us now clarify this dissimilarity between the photon and the wave packet from yet another point of view. The former concept arises in the quantization of the field. Thus, if $\hat{a}_k{}^+$ and $\hat{a}_k$ are the operators which, respectively, create and destroy a quantum of energy $\hbar\omega_k$ and wave vector $\boldsymbol{k}$, then these quanta are known as the photons [*102*]. The number of photons with the wave vector $\boldsymbol{k}$ excited in the cavity is determined by the eigenvalue s_k of the appropriate number operator $\hat{s}_k \equiv \hat{a}_k{}^+\hat{a}_k$. It is to be noted that the notion of the energy exchange is of prime importance in this definition. In addition, the value of s_k will vary depending on the surrounding medium.

The wave packet, on the other hand, is defined as being the result of a superposition of nearly monochromatic waves and is seen as occupying some finite region in space [50]. It is described mathematically by

$$A(x, t) = \frac{1}{\sqrt{2\pi}} \int_{-\infty}^{\infty} A(k)\, e^{i(\boldsymbol{k}\cdot\boldsymbol{r}-\omega t)} dk\,, \tag{5.1}$$

where it must be remembered that ω is a function of $\boldsymbol{k}$. The harmonic term contains information about the velocity of propagation of phase and it, therefore, belongs to the undulatory domain.

Next, the momentum density of an elementary radiation field is given by Abraham as

$$g_A = \frac{h\nu}{cn} \tag{5.2}$$

and by Minkowski as

$$g_M = \frac{nh\nu}{c}. \tag{5.3}$$

In a plasma, where $n < 1$, it holds that

$$g_M < g_0 < g_A\,, \tag{5.4}$$

where $g_0 = h\nu/c$. One is, therefore, led to suggest that, in a plasma, g_M refers to a "bare" and g_A a "dressed" particle, since a "bare" particle influenced by (i.e., interacting with) a medium is referred to as a "dressed" one. As a particle has, in general, a medium-dependent effective mass, only that latter kind is of experimental significance. Hence, it is proposed that the "bare" particle be associated with the concept of a photon, whereas the "dressed" one be linked with that of a wave packet. Putting it in another way, the radiation, when interfering with matter, assumes some of the characteristics of the latter. Thus, the initially pure, electromagnetic wave becomes now modulated by the surrounding medium. However, when a photon is absorbed by a constituent of matter, the wave packet description combines the properties of the material constituent with those of the absorbed photon.

The above discussion distinguishes between the complementary meanings of the photon and the wave packet for radiation in a medium. It is, hence, conjectured that the momentum density g_M applies to a photon and g_A to a wave packet, whenever an electromagnetic field inside a medium with the refractive index less than unity, is considered.

The above proposition is in agreement with the work of Hora [39, 103] and Klima and Petrzilka [82], who found the total radiation momentum in a plasma medium to be given by

$$g = \frac{1}{2}\,(g_A + g_M). \tag{5.5}$$

This represents, of course, the average of the two momenta: that of the photon and that of the wave packet. In addition, if the last term in the Peierls expression (3.62), which arises from the presence of dipoles, is neglected, then the total momentum is again given by (5.5). Although this interpretation is somewhat arbitrary, it, nevertheless, leads to satisfactory conclusions.

Consider now radiation propagating in a medium and assume the conditions to be such that laws of geometrical optics apply. This means that the wavelength is much smaller than the typical dimensions involved in the problem. In this case, therefore, waves are treated as corpuscules travelling along the ray trajectories, completely lacking in any undulatory characteristics. The role of phase and group speeds in this asymptotic picture becomes clearer from the following considerations, due to PARK [*104*]. The scalar d'Alembert wave equation

$$\Delta\psi + k^2\psi = 0; \qquad k = \frac{\omega}{v} \tag{5.6}$$

degenerates in the limit of $\lambda \to 0$. Let us assume a solution of (5.6) to have the form

$$\psi = ae^{iS}, \tag{5.7}$$

where S is the eikonal. In such a case, (5.6) will be satisfied by

$$(\nabla S)^2 = \frac{\omega^2}{c^2} n^2. \tag{5.8}$$

This, in turn, means that the surfaces $S =$ const. describe surfaces of constant phase and, consequently, determine the phase velocity.

On the other hand, the Hamilton-Jacobi eikonal

$$(\nabla S)^2 = 2m\big(E - V(x)\big) \tag{5.9}$$

determines the group velocity which is necessary in order to find the time a particle spends in an optical region. This is connected with the work of LEWIS [*105*], who divided the transport equations into two groups: those describing intensity and those referring to the phase change.

A different approach is adopted here. By extending the Fresnel theorem of stationary phase, DE BROGLIE [*106*] arrived at the relation between the speed v of a massive particle and the phase velocity v_p of the corresponding wave

$$vv_p = c^2. \tag{5.10}$$

Furthermore, he found that the group speed of the wave, associated with a particle, is equal to the particle speed. Relation (5.10), known as the Broglie group velocity theorem, may be interpreted as pointing to some form of connection in the particle and phase velocities. Of course, this theorem also applies to particles with vanishing rest mass. Then, it expresses the constancy of the product of the corpuscular and undulatory speeds. However, as the energy transport decreases in a medium, the separation into those two aspects is more obvious.

In view of (5.10), it should be remarked that

$$g_M \cdot g_A = g_0{}^2, \tag{5.11}$$

which is a relation connecting the Minkowski, Abraham, and vacuum values of the momentum densities. Therefore, (5.11) can be looked upon as describing the link between the corpuscular and undulatory aspects of the momentum. Although a direct physical significance cannot be attached to the undulatory momentum, this need be no cause for concern, as this analysis applies only in the WKB (i.e., effectively geometrical optics) approximation. It will be seen below that the Minkowski value of the momentum (representing photons) does, indeed, satisfy the condition of equilibrium between radiation and a plasma (or, generally, a medium with $n < 1$), whereas, as expected, the Abraham value does not.

The conservation of momentum and center of mass theorems were already discussed in Chapter III. They were applied there to the analysis of the photon momentum inside an optically dense ($n > 1$) medium. Although that approach is rather simplistic, it will be used here to cover also the domain of optically thin ($n < 1$) media. The qualitative analysis is, essentially, the same as before and, hence, it is not presented here. Nevertheless, it is to be noted that only the Minkowski result simultaneously satisfies both the conservation of momentum and the velocity invariance theorems.
There is some connection between the above approach and that discussed by FOCK [*107*] who draws analogy, from the purely formal standpoint, between classical particles and optical waves. Furthermore, HORA [*109*] and CARTER and HORA [*110*] applied this similarity to investigate the Goos-Hänchen effect for matter waves. In addition, arguments presented by JOYCE [*76, 77*] attempt to reformulate classical mechanics so as to encompass particles of vanishing as well as non-vanishing rest masses. An application of matter waves to investigate the propagation of light in fibres was presented by ARNAUD [*108*]. To illustrate the similarity between those two types of waves, he referred to the quantum mechanical AHARONOV-BOHM effect [*111, 112*], which has been experimentally verified [*113*]. The classical explanation of this effect was given by BOYER [*114*]. Basically, the AHARONOV-BOHM effect predicts the physical significance of the electromagnetic potentials, and also suggests that their role, in quantum mechanics, is similar to that of the refractive index in classical physics. To verify this proposal, a simple experimental set up is used to observe diffractive effects between matter and light waves.
Thus, consider a row of metallic channels which are at gradually increasing potentials. A plane matter wave (such as that corresponding to electrons) entering those channels at one end, will suffer a deflection upon exit, despite the absence of any classical forces. Similarly, a light wave passing through an optical phase array (a row of channels with increasing optical densities) will be also deflected. Moreover, if the dispersion of the light waves inside this optical system is the same as that for the matter waves in vacuo, then the resultant deflection of the wave will be equal.
This correspondence between matter and optical waves will be further elaborated in the next chapter. Now, let us return to the question of the electromagnetic momentum in a plasma and investigate the important topic of the black body radiation. That last phenomenon describes, as is well known, the equilibrium spectral distribution between the thermal radiation and matter in an evacuated cavity. The presence of matter is necessary to satisfy the equilibrium conditions and its amount is minute. When, however, the cavity is filled with a plasma, it affects this spectral distribution and it is quite natural to suppose that the density of radiation inside plasma differs from the one in vacuo. Thus, if $\varepsilon(\nu, T)$ is the mean energy of oscillators of frequency ν at the temperature T, then

$$\varepsilon(\nu, T) = h\nu(e^{h\nu/\varkappa T} - 1)^{-1} \tag{5.12}$$

and the energy density $u(\nu)$ of the radiation can be expressed by

$$u(\nu)\,d\nu = \varepsilon(\nu, T)\,g(\nu)\,d\nu, \tag{5.13}$$

where $g(\nu)\,d\nu$ refers to the density of radiation states in the frequency interval ν to $\nu + d\nu$. This density of states is usually defined, in terms of the wave vector $\boldsymbol{k}$, as

$$g(\boldsymbol{k})\,d\boldsymbol{k} = \pi^{-2}k^2 d\boldsymbol{k}. \tag{5.14}$$

In order to arrive at the density of states, in terms of frequencies, (5.14) is transformed, utilizing (4.36), into

$$g(\nu)\,d\nu = \frac{8\pi}{c^3}\,n\nu^2 d\nu. \tag{5.15}$$

This last expression is in accordance with the one obtained by DAWSON and OBERMAN [*115*] and, if the group refractive index is set equal to the usual phase refractive index, then the corresponding relation in the work of OSTER [*116*] reduces to (5.15).
Now, if (5.15), together with (5.12), is inserted into (5.13), we obtain the following modified Planck law for an isotropic, collisionless plasma

$$u(\nu) = n u_0(\nu), \tag{5.16}$$

where $u_0(\nu)$ corresponds to the vacuum value of the energy density. The result (5.16) was obtained in the literature by a completely different method by BEKEFI [*36*] and DAWSON [*37*]. CASE and CHIU [*117*] used the fluctuation-dissipation theorem to arrive at the same conclusion. Although OSTER [*116*] also obtained (5.16), his radiative transfer theory was criticized by CRONYN [*118*] as violating the Second Law of thermodynamics. In addition, the transport equation [*116*] differs from that due to MARTYN [*119*]. A modified form of the Planck law in anisotropic media was described by COLE [*38*], which, in the isotropic limit, reduces to (5.16).
Neglecting the longitudinal waves, INMAN [*120*] developed the thermodynamics for electromagnetic waves in plasma. He found that, as the plasma frequency increases, the number-, free energy-, entropy- and energy-densities are all decreasing from their vacuum values. The correctness of the formula for the radiation pressure given by Inman was, however, disputed by MORE [*121*].
We are now in a position to develop further the method used by EINSTEIN [*34*] in his investigation of emission and absorption processes. An attempt was made by SKOBEL'-TSYN [*35*] to apply this method to optically dense media. His work, however, appears to be inconclusive. In the present treatment, the original (Einstein) method will be adapted to the plasma domain.
During any absorption (or emission) process, a momentum is transferred to an elementary system in the plasma. The velocity distributions of such systems are merely due to their interactions with the radiation field. Thus, this distribution must be the same as the one these systems attain through their mutual collisions alone. Hence, it is necessary that the average kinetic energy of a system in the radiation field be equal to $(1/2)\varkappa T$.
Suppose now that there are two possible quantum states z_m and z_n with the corresponding energies ε_m and ε_n. Next, let the transition from state z_m to state z_n be allowed for the system. That transition will then result if the system in state z_m absorbs a quantum of energy of magnitude $\varepsilon_n - \varepsilon_m$. Similarly, the transition from z_n to z_m will result when the system emits that quantum of energy. Now, there are two different processes a system can undergo to make a transition. First, there is a process independent of the external radiation field, viz., the so called spontaneous emission. The probability for such a process to occur in time dt is

$$dP = A_{nm}\, dt, \tag{5.17}$$

where A_{nm} is the Einstein "A" coefficient. Second, a transition which depends on the stimulation by the external field. Processes of this types are known as "Induced" radiation processes. Thus, the probability dP that a system under the influence of the radiation density $u(\nu)$ makes the transition from z_m to z_n by absorbing the energy quantum is $dP = B_{mn} u(\nu)\, dt$. Similarly, the probability for the transition from z_n to z_m to occur is $dP = B_{nm} u(\nu)\, dt$. Again, B_{mn} and B_{nm} are the Einstein "B" coefficients. Thus, in those two cases, the transition probability is the function of the radiation energy density $u(\nu)$ surrounding our elementary system. In other words, as the system is in a plasma, $u(\nu)$ refers to the radiation density inside the latter. This density is given by (5.16) and its use is imperative. (That this choice does not disturb the thermodynamic equilibrium conditions will be shown below.) Never the less, SKOBEL'TSYN [*35*] originally

assumed that the radiation energy density in a dielectric is the same as the one in vacuo, but, in the same article, he later made an ad hoc assumption that that density ought to be $u(\nu)/n$, for $n > 1$.
In view of the above discussion, therefore, the probabilities for upward and downward transitions can now be written, respectively, as

$$dP = nB_{mn}u_0(\nu)\,dt; \qquad dP = nB_{nm}u_0(\nu)\,dt. \tag{5.18}$$

Hence, the probability for the radiation induced processes in a collisionless plasma is decreased relative to vacuo. This is so, because, as the plasma density increases, the refractive index decreases and, thus, the amount of radiation energy density present is diminished, in accordance with the expression (5.16).
Investigate now the motion of a test particle under the influence of the radiation field. To simplify the calculations and to make, at the same time, the physics of the problem more transparent, it will be assumed that the particle of mass M is moving with velocity v in the x-direction and that the conditions are such that the higher powers of v/c can be neglected. In a radiation field, the linear momentum Mv of this particle changes, in time τ, due to the effect of the radiation force $f = -Rv$, R being constant to be evaluated below. Also, during the same time, the fluctuations of the action of radiation on the particle cause a momentum transfer Δ to the latter. The magnitude and the direction of this momentum transfer are constantly changing. As the radiation field must not affect the overall velocity distribution, the average momentum of the particle, at both ends of the time interval τ, must be $\overline{Mv}$. Therefore, we must have

$$\overline{(Mv - Rv\tau + \Delta)^2} = \overline{(Mv)^2}. \tag{5.19}$$

We shall work in the approximation in which all terms containing $\overline{v\Delta}$ can be neglected. Further, in order not to disturb the thermodynamic equilibrium, we shall assume the relation $(1/2)\overline{Mv} = (1/2)\varkappa T$ to be valid. The above equation then reduces to

$$\frac{\overline{\Delta^2}}{\tau} = 2R\varkappa T. \tag{5.20}$$

Thus, this is the fundamental equation required by the theory of heat and a comparison with [*34*] affirms that the presence of a plasma does not affect it.
Now, if $\overline{\Delta^2}/\tau$ and R are calculated and then inserted in (5.20), it must lead to an identity, as has been already pointed out by Einstein [*34*]. However, in order to calculate these quantities using his method, in the conditions of thermal equilibrium, it is necessary to know the radiation energy density and the photon momentum. This, in turn, requires the knowledge of the field energy momentum tensor. It, therefore, follows that the condition (5.20) can be used to ascertain the correct form of this tensor.
As the radiation laws are defined only for those inertial frames in which our particle is at rest, in calculating constant R, the transformation formulae [*2*] of the special theory of relativity are used. Further, the quantities referring to the rest frame of the particle will be marked with a dash. Thus, radiation per unit volume in the frequency range $d\nu$, within a solid angle $d\Omega$, transforms according to

$$u'(\nu', \phi') = u(\nu)\,\frac{d\nu}{d\nu'}\,\frac{d\Omega}{d\Omega'}\,(1 - 2n\beta\cos\phi), \tag{5.21}$$

where $\beta = v/c$ and $u'(\nu', \phi')$ depends on the direction defined by the angle ϕ' it makes with the x'-axis, and the angle ψ' between its projection in the $y'z'$-plane and the y'-axis.

To facilitate the transformation of the argument of $u'(\nu', \phi')$, the formulae for the Doppler effect and the aberration are needed. These, however, are well known in the theory of relativity and, in the approximation applicable here, take the forms

$$\nu' = \nu(1 - n\beta \cos \phi) \tag{5.22}$$

and

$$\cos \phi' = \cos \phi - n\beta + n\beta \cos^2 \phi , \tag{5.23}$$

respectively. Further, no relevant information is lost if we put

$$\psi' = \psi . \tag{5.24}$$

The inverse transformation to that in (5.22) is

$$\nu = \nu'(1 + n\beta \cos \phi') \tag{5.25}$$

which, on substitution in $u(\nu)$ and subsequent expansion, gives

$$u(\nu) = u(\nu') + \nu' n\beta \cos \phi' \frac{\partial u(\nu')}{\partial \nu'} . \tag{5.26}$$

The solid angle $d\Omega$ is (in our approximation) connected with $d\Omega'$ by

$$d\Omega/d\Omega' = 1 - 2n\beta \cos \phi' . \tag{5.27}$$

Thus, the substitution of (5.25), (5.26), and (5.27) in (5.21) results in

$$u'(\nu', \phi') = \left[u(\nu') + \nu' n\beta \cos \phi' \frac{\partial u(\nu')}{\partial \nu'}\right] (1 - 3n\beta \cos \phi') . \tag{5.28}$$

Making use of the last relation and the assumption about the elementary processes, discussed earlier, allows calculation of the average momentum transferred, per unit time, to the particle. Let, then, the number of elementary transitions from level m to level n be

$$\frac{1}{S} w_m e^{-\varepsilon_m/\varkappa T} B_{mn} u'(\nu', \phi') \, d\Omega'/4\pi , \tag{5.29}$$

where

$$S = w_m e^{-\varepsilon_m/\varkappa T} + w_n e^{-\varepsilon_n/\varkappa T}$$

and w_m, w_n are the statistical weights corresponding, respectively, to levels m and n. An analogous form holds for the transition from level n to level m.

As suggested earlier, the photon momentum in plasma is given by the Minkowski relation. Therefore, in each elementary process, the momentum $nh\nu \cos \phi'/c$ is transferred to a particle. Combining the upward and downward transitions, the total momentum transferred to the particle is then expressed, using (5.29), as

$$nh\nu(Sc)^{-1} w_m B_{mn} (e^{-\varepsilon_m/\varkappa T} - e^{-\varepsilon_n/\varkappa T}) \int u'(\nu', \phi') \cos \phi' d\Omega'/4\pi . \tag{5.30}$$

Substituting (5.28) into (5.30) and putting $Q = (1/S) \, w_m B_{mn} e^{-\varepsilon_m/\varkappa T}$, the total averaged momentum transferred, per unit time, to a particle through the induced processes is

$$-\frac{n^2 h\nu}{c} \beta \left[u(\nu) - \frac{\nu}{3} \frac{\partial u(\nu)}{\partial \nu}\right] Q(1 - e^{-h\nu/\varkappa T}) . \tag{5.31}$$

The dashes over symbols have been dropped here, for the sake of clarity.

As the spontaneous emissions do not have any preferred directions, average momentum transfers, due to these processes, vanish.
Thus, finally, the coefficient R can be written as

$$R = \frac{n^2 h\nu}{c^2} \left[u(\nu) - \frac{\nu}{3} \frac{\partial u(\nu)}{\partial \nu} \right] Q(1 - e^{-h\nu/\varkappa T}). \tag{5.32}$$

Prior to utilizing this result, let us inquire about the effect of the presence of a plasma on the momentum Δ. Denote by δ the momentum transferred to the particle during some process and assume that its sign and magnitude vary in time in such a way that the average value of δ vanishes. Now, if δ_i refers to the transfer of momentum through the i-th independent process, then the resultant transfer is $\Delta = \sum \delta_i$. Because the average value of each δ_i vanishes, it is necessary to write (assuming $\overline{\delta_i^2} = \overline{\delta^2}$)

$$\overline{\Delta^2} = N\overline{\delta^2}, \tag{5.33}$$

where N is the number of all elementary transitions in time τ. If the momentum corresponding to each elementary process, $\delta = nh\nu \cos\phi/c$, is substituted into (5.33), it then yields

$$\overline{\Delta^2} = \frac{N}{3} \left(\frac{nh\nu}{c} \right)^2. \tag{5.34}$$

As the number of induced processes from level m to level n, in time τ, in equilibrium, is $N/2$, it is, then, possible to write

$$N = 2Qu(\nu)\,\tau,$$

which, on combining with (5.34), gives

$$\frac{\overline{\Delta^2}}{\tau} = \frac{2}{3} \left(\frac{nh\nu}{c} \right)^2 Qu(\nu). \tag{5.35}$$

As already previously emphasized, the expressions (5.32) and (5.35) are determined in terms of the radiation energy density $u(\nu)$ which, of course, differs from its vacuum value $u_0(\nu)$. An attempt will now be made to show the validity of the claim that the photon momentum inside plasma is correctly described by the Minkowski relation. For that purpose, we must show that the thermodynamic equilibrium is not disturbed when the radiation momentum is transferred to a particle. This, in turn, will be so if, after calculating R and $\overline{\Delta^2}$ in terms of the proposed energy and momentum densities of radiation, their consequent insertion in (5.20) yields an identity.
Thus, substitution of (5.16) in (5.32) results in

$$R = \frac{nh\nu}{c^2} \frac{u_0(\nu)}{3} \left(n^2 \frac{h\nu}{\varkappa T} \frac{e^{h\nu/\varkappa T}}{e^{h\nu/\varkappa T} - 1} + n^2 - 1 \right) Q(1 - e^{-h\nu/\varkappa T}). \tag{5.36}$$

Now, if conditions are such that $h\nu/\varkappa T \gg 1$, then (5.36) reduces to

$$R = \frac{n^3 h\nu}{c^2} \frac{u_0(\nu)}{3} \frac{h\nu}{\varkappa T} Q. \tag{5.36a}$$

Furthermore, when (5.16) is inserted in (5.35), the result is

$$\frac{\overline{\Delta^2}}{\tau} = \frac{2}{3} n^3 \left(\frac{h\nu}{c} \right)^2 u_0(\nu)\, Q. \tag{5.37}$$

Finally, if (5.36a) and (5.37) are substituted into (5.20), an identity relation is, indeed, obtained. It can be, therefore concluded that the Minkowski form of the momentum density, used in the preceding analysis, appears to be satisfactory, at least in the region of high frequencies and within a collisionless plasma. On the other hand, the use of the Abraham momentum density does not satisfy (5.20) and, therefore, must be considered inappropriate.

When interparticle collisions are present in a plasma, the above analysis should still apply, provided that we include the effect of those collisions in our calculations. One possible way of doing this is to introduce a complex refractive index (4.26), which then takes into account the absorption due to collisions. It is, then, sufficient, for $\nu_{\text{eff}} \ll \omega$, to replace the value of the refractive index in the above relations by the corresponding magnitude of its complex counterpart. This is, essentially, the approach used by HORA [*39*] whilst treating the interaction of laser radiation with an inhomogeneous plasma. In this chapter, results, valid in what is, essentially, a WKB approximation, were presented. In the following pages, a more general approach will be adopted to facilitate our inquiry into the form of the electromagnetic energy-momentum tensor in a plasma.

VI. The Proca Equations

Results obtained in the previous chapter should be understood as the asymptotic solutions to the wave optics viewed from the classical standpoint. It appears, in this light, that the only expression for the momentum exchange satisfying (5.20), the momentum exchange between field and a particle, is the Minkowski one. It is to be noted, however, that no generally applicable conclusion about the energy-momentum tensor can be reached through utilizing the method of the preceding chapter, as that formulation applies only in the case of geometrical optics. Furthermore, it is not relativistically covariant.

In order to arrive at a reliable result, some other, more rigorous, method must be used. Thus, the procedure to be presented here, resorts to the formalism of the covariant field theory, which encompasses the realm of wave phenomena, as well as the corpuscular approach, just presented.

Quite generally, a medium, in which the phase speed is frequency dependent, is known as a dispersive one. In such a case, the relation between the angular frequency ω of the wave and the corresponding wave vector $\boldsymbol{k}$ assumes the form [*50*]

$$\omega^2 = k^2c^2 + \omega_p{}^2\eta^2, \tag{6.1}$$

where

$$\eta^2 = \frac{\omega^2}{\omega^2 - \omega_0{}^2}$$

and ω_p is the plasma frequency, whilst ω_0 describes the characteristic excitation frequency of the medium. In a plasma, ω_0 vanishes. The expression (6.1) is only approximate, as it is derived under the assumption that the dispersion is determined solely by electrons.

In the special theory of relativity, the relation between the energy E and the momentum $\boldsymbol{p}$ of a particle, having rest mass m_0, is given by

$$E^2 = p^2c^2 + m_0{}^2c^4. \tag{6.2}$$

A quick comparison of (6.1) with (6.2) reveals a striking similarity between both relations. Thus, it was suggested [*40, 41*] that the propagation of a wave in a medium may be treated analogously to the motion of a massive particle in vacuo. Put otherwise, as the wave propagates through a collisionless plasma [43], collisional plasma [122], or a nondispersive fluid [83], it interacts with its surroundings and, thus, takes on some of the properties of the medium. Thus, for instance, the effect of a collisionless plasma on the field is contained in the ω_p term in (6.1), since, in that case, $\eta = 1$. Hence, a photon which, before entering a plasma region, had a zero rest mass, acquires, whilst travelling inside it, an effective rest mass of $\hbar\omega_p/c^2$. This is readily obtained from (6.1) and (6.2). Although it is normally assumed that photons are massless particles, this requirement is not necessary from the point of view of the special theory of relativity. What is really important, is that c, the speed of light in vacuo, remains constant in moving inertial frames. This, however, would point to a particle with a vanishing rest mass. On the other hand, the speed of a massive photon must be less than c and ought to depend on its energy, $h\nu$, and, hence, its frequency. This would imply that electromagnetic waves of different frequencies would propagate with different speeds. In fact, it has been suggested that high frequency waves should propagate at higher speeds than the low frequency ones. As the waves of progressively lower frequencies are considered, the limit is eventually reached when the propagation speed drops to zero. This happens when $\hbar\omega = m_0c^2$, i.e., in view of the above discussion, when the frequency of the incident wave is equal to the plasma frequency. Experimental results [123] obtained from pulsars agree with the just presented interpretation.

Therefore, it seems quite plausible to replace an electromagnetic wave in a medium by an equivalent massive photon in vacuo. It will be recalled that the principal difficulty in establishing the correct form of the energy-momentum tensor in a macroscopic body lay in the impossibility of splitting that tensor uniquely into the field and the matter parts. That separation becomes possible here, if we are willing to accept the above interpretation.

The consequences of replacing propagation of radiation in a medium by motion of a photon with a nonvanishing rest mass are rather intriguing. Thus, it follows from the Maxwell theory that the photon can be polarized in two directions. However, as the photon has now an effective mass, the generalization of the Maxwell equations leads to the theory of Proca equations. Consequently, there should occur now all three states of polarization which are predicted by the latter theory. Thence, apart from two transverse polarizations, there also exists a longitudinal one. Furthermore, the faster the longitudinal photon moves, the weaker is the associated electric field [124], thus, implying only a rather weak interaction with a medium. In addition, the electric field diminishes exponentially with distance [124] and the corresponding flux lines fade away, even in vacuo. Similarly, magnetic field is also affected and the associated lines are compressed around equator.

Using methods of classical electrodynamics [50], we proceed now to analyze this problem more rigorously. Although there will be some overlap with Chapter II, as far as the presentation is concerned, the conclusions we reach here will be, in general, quite different.

It is necessary, at this stage, to discuss the problem of the photon spin. Thus, according to quantum mechanics, the total angular momentum of the radiation field is equal [125] to the sum of the spin and the orbital angular momenta. The spin of the photon is, generally, accepted to be unity and to lie in the direction of propagation. This was verified by observation [89]. We shall, therefore, assume here that the spin of the massive photon is also unity. Hence, in view of the above discussion, the Maxwell field in a medium is, essentially, replaced by a neutral vector meson field in vacuo. Then, it is applied to the question of the energy-momentum tensor.

Thus, a free electromagnetic wave corresponding to a massive photon may be described by

$$\boldsymbol{A} = \boldsymbol{A}_0 e^{-i(\omega t - \boldsymbol{k}\cdot\boldsymbol{r})} \tag{6.3}$$

with

$$\left(\frac{\omega}{c}\right)^2 - k^2 = \mu^2, \tag{6.4}$$

where μ is the photon's rest mass, in units of inverse length. From its definition, $v_p = \omega/k$, the phase speed is

$$v_p = c\left(1 + \frac{\mu^2}{k^2}\right)^{1/2}, \tag{6.5}$$

whilst, for the group speed, $v_g = d\omega/dk$, we have the expression

$$v_g = \frac{c}{\omega}(\omega^2 - \mu^2 c^2)^{1/2}. \tag{6.6}$$

A glance at (6.6) shows that also for such a massive field, the dependence of the energy propagation on frequency is in accordance with experimental findings [*123*, *124*].
Let us consider, for a moment, the case of a static field, i.e., $\omega = 0$. It then follows from (6.4) that $k = \pm i\mu$. Consequent substitution into the wave equation (6.3) implies that the spherical static field decays exponentially, i.e., $A = A_0 e^{-\mu r}(A_0 \propto 1/r)$ as already indicated. On selecting $k = i\mu$, we have a behaviour similar to that displayed by the Yukawa, model of nuclear binding. Thus, as one of the consequences of the photon being massive, we must have a deviation from the Coulomb law. Hence, in view of our supposition, the field in a collisionless, isotropic plasma should decay exponentially with distance. The same conclusion was also obtained by COLE [*40*], in his study of wave propagation in an inhomogeneous medium. He replaced ω by $i\partial_t$ and $\boldsymbol{k}$ by $-\partial_i$ in the dispersion relation and arrived at the Klein-Gordon equation

$$c^2\partial_i^2\phi - \partial_t^2\phi = \omega_p^2\phi. \tag{6.7}$$

In the case of a spherically symmetric static field, the solution of this is clearly

$$\phi = \frac{q}{r} e^{-r\omega_p/c}. \tag{6.8}$$

This Yukawa type behaviour applies to the case of an isotropic plasma, in view of the form of the dispersion relation (6.1). On introducing, in addition, a magnetic field, the solution of the Klein-Gordon equation, in the static limit, takes the form of the Coulomb potential, thus, effectively, removing the shielding effect of c/ω_p.
There is another interesting point reported by COLE [*40*]. He noticed that the parameter c/ω_p also occurs in the solid state literature, where it is referred to as the London penetration depth. However, as we saw above, it also distinguishes a plasma without a magnetic field from one with the field. In other words, it distinguishes between superconducting and conducting states.
As the photon now is assumed to have a nonvanishing effective rest mass, a mass term has to be added to the Lagrangian density $\mathcal{L}$. The thus obtained Lagrangian becomes [*50*]

$$\mathcal{L}_p = -\frac{1}{16\pi} F_{\alpha\beta}F^{\alpha\beta} - \frac{1}{c} j_\alpha A^\alpha + \frac{\mu^2}{8\pi} A_\alpha A^\alpha \tag{6.9}$$

and is termed the Proca Lagrangian. The reciprocal length in (6.9) is associated with the rest mass of the photon by $\mu = m_0 c/\hbar$. The Proca equations of motion are then written as (cf. 2.9)

$$\partial_\mu F^{\mu\nu} + \mu^2 A^\nu = \frac{4\pi}{c} j^\nu, \tag{6.10}$$

where $F^{\mu\nu}$, the electromagnetic field tensor, is given by (2.4). Thus, (6.10), together with (2.4), gives us the covariant Proca equations for a massive vector field, and is should be noted that it is the only possible linear generalization of the Maxwell equations [126].
It is to be noted that, as a result of the inclusion of the coupling constant μ, the potentials now acquire real physical characteristics. In other words, in contradistinction to the potentials in the Maxwell theory, they now become observable [127].
If we, further, assume the validity of the Lorentz gauge, $\partial_\alpha A^\alpha = 0$, then (6.10) may be rewritten as

$$\Box A^\mu + \mu^2 A^\mu = \frac{4\pi}{c} j^\mu, \tag{6.11}$$

where the four-dimensional Laplacian operator (also known as the d'Alembertian) is defined by $\Box \equiv \partial_\alpha \partial^\alpha$. When time independence of A^μ is assumed, so that the field may be regarded as static, the solution of (6.11) leads to the Yukawa expression for the potential, as already discussed.
In three dimensional notation, the modified Maxwell equations are [126]

$$\begin{gathered}\nabla \times \boldsymbol{E} = -\frac{1}{c}\,\partial_t \boldsymbol{H}, \qquad \nabla \times \boldsymbol{H} = \frac{1}{c}\,\partial_t \boldsymbol{E} + \frac{4\pi}{c}\boldsymbol{j} - \mu^2 \boldsymbol{A} \\ \nabla \cdot \boldsymbol{E} = 4\pi\varrho - \mu^2 V, \qquad \nabla \cdot \boldsymbol{H} = 0 \\ \frac{1}{c}\,\partial_t V + \nabla \cdot \boldsymbol{A} = 0, \qquad \nabla \times \boldsymbol{A} = \mu^2 \boldsymbol{H}, \quad \nabla V + \frac{1}{c}\,\partial_t \boldsymbol{A} = -\mu^2 \boldsymbol{E}.\end{gathered} \tag{6.12}$$

These comprise the set of fifteen Proca equations and may be viewed as forming two physically distinct groups [127]. Thus, four top equations represent, apart from the coupling constant μ, the Maxwell field through the six vector ($\boldsymbol{E}, \boldsymbol{H}$); the bottom three equations, on the other hand, describe a four-gradient of a scalar potential. A closer inspection reveals that those groups may be seen as corresponding to the transverse and longitudinal waves, respectively.
It is generally accepted that, in the Maxwell model, the vectors $\boldsymbol{E}$, $\boldsymbol{H}$, and $\boldsymbol{k}$ are all mutually orthogonal. Furthermore, the scalar potential may be made to vanish there, so that the vector potential $\boldsymbol{A}$ is parallel to $\boldsymbol{E}$. Assuming that the same should apply also in the Proca theory, for fields corresponding to transverse polarizations, we obtain from the Lorentz transformation relations [2] that [127]

$$\boldsymbol{E} = -\frac{\omega}{\mu c}\boldsymbol{A} \qquad \frac{|\boldsymbol{H}|}{|\boldsymbol{E}|} = \frac{ck}{\omega}. \tag{6.13}$$

Similarly, in the case of longitudinal waves, the vectors $\boldsymbol{E}$, $\boldsymbol{A}$, and $\boldsymbol{k}$ are mutually parallel and, further, $\boldsymbol{H} = 0$. Therefore, the following relations hold

$$\boldsymbol{E} = -\frac{\mu c}{\omega}\boldsymbol{A} \qquad V = \frac{ck}{\omega}\boldsymbol{A}. \tag{6.14}$$

Now, utilizing the dispersion relation, $\omega^2 - k^2c^2 = \mu^2c^2$, it becomes clear that the four-vector (ω, $\boldsymbol{k}c$) must be time-like. Hence, it is possible to transform it to the rest frame of the massive photon. (In the case of a massless photon this transformation, of course, fails, as the four-vector is then light-like.) Thence, in the heavy photon's rest frame $\boldsymbol{k} = 0$. Consequently, (6.13) yields $\boldsymbol{H} = 0$ and (6.14) gives $V = 0$. Because, in both cases, the vectors $\boldsymbol{E}$ and $\boldsymbol{A}$ are parallel, the resulting $\boldsymbol{E}$-field is just the expression of mutual transformability between transverse and longitudinal waves. This, in turn, is a direct consequence of the fact that the state of polarization is not the Lorentz invariant [124].

Moreover, the longitudinal wave propagates in the direction of the common wave vector with the velocity equal to that of the transverse wave [*127*]. Longitudinal and transverse waves must be, therefore, treated as being equally important. However, this appears to lead to the conclusion that an extra degree of freedom must be taken into account in discussing the black body radiation. In fact, as a consequence, a factor of 3/2 should now appear in the Planck formula. But, as explained in [*127*], the transition between the Maxwell and Proca fields must be continuous. In other words, the factor 3/2, which would introduce a clear discontinuity between the two fields must be spurious and should not appear whenever a transition is made between massless and massive photons.

Consider now a vector field in a cavity with perfectly reflecting walls. Applying the Fresnel formulae [*50*] gives [*127*] a transverse wave which is nearly completely reflected but contains, after reflection, a small fraction of a longitudinal wave. A small amount of energy is, at the same time, carried away by the transmitted longitudinal wave. The effect of a longitudinal wave on the matter in its path is very small [*128*], so that virtually no transfer of momentum to the obstacle occurs. Because the electric field is so small (according to (6.14), it is proportional to $\mu c/\omega$) the interaction occurring between this type of wave and matter can be, for most purposes, neglected. In other words, the approach of the longitudinal photon to equilibrium is very slow. Thus, for example, assuming the rest mass of a photon $m_0 = 10^{-38}$ g and an angular frequency ω in the optical region in a physical cavity, it can be estimated [*127, 128*] that the time necessary for the transverse photons to transform into longitudinal ones is of the order of 10^{11} years. This transformation effect is, indeed, minute.

Furthermore, in order to avoid the contradiction with the black body radiation law, GOLDHABER and NIETO [*124*] suggested that the longitudinal photons may have a rest mass of several GeV so that the interaction length is so small that the effect cannot be, essentially, observed. Their reasoning follows from the rule known in particle physics which states that the range of a propagated interaction is inversely proportional to the mass of the particle responsible for its transmission.

GERTSENSHTEIN and SOLOVEI [*129*], taking into account that even the best conductors are not capable of reflecting these waves, introduced the concept of the skin depth δ_L, corresponding to the longitudinal waves. This is related to the skin depth δ_T for the transverse waves by

$$\delta_L \geqq (\omega/\mu c)^2 \, \delta_T, \tag{6.15}$$

where the equality holds for low conductivity. Due to the factor $(\omega/\mu c)^2$, the transparency of the stellar medium (which is in plasma state) to the longitudinal photons is nearly perfect, thus, explaining the lack of their observation. It was, further, pointed out by WIGNER (cf. [*124*]) that the longitudinal waves cannot be responsible for cooling of the stars, as the rate of their production must be minute. He then concluded that the thermodynamics of a system in question would be barely affected by the presence of the longitudinal photons.

It is important to emphasize, at this point, that although the above mentioned workers assumed the photon mass to be non-vanishing in vacuo, the underlying philosophy of the present work is diametrically opposite. We assume the standard Maxwell field in vacuo, implying a zero rest mass of a photon in empty space. Only the study of fields inside matter requires an introduction of massive photon electrodynamics. In that case then, the coupled (Maxwell) field is replaced by the free (Proca) field. Put otherwise, the analysis of the behaviour of a photon in a medium, where the dispersion relation, $\omega^2 = k^2c^2 + \omega_p{}^2$, holds, is replaced by investigation of the properties of a neutral vector meson in vacuo.

Next, let us apply this interpretation to some of the available experimental observations.

It is generally accepted that by far the largest part of the Universe is in a plasma state. Therefore, it seems quite natural to turn our attention towards cosmic phenomena. From the definition of the refractive index, $n^2 = 1 - (\omega_p^2/\omega^2)$, it, clearly, follows that values of n at radio frequencies are much smaller than those at optical frequencies. Thus, the variation of propagation of an electromagnetic wave through plasma, is, evidently, greatest at small frequencies. Therefore, for small electron densities, radio waves should yield more pronounced effects. As the interstellar electron densities are not known very accurately [*33*], there is some uncertainty in the effective mass of the photon. With this in mind, COLE [*40*] arrived at the following values: in ionosphere $m_0 \approx 10^{-40}$ g, whilst in solid state plasma $m_0 \simeq 10^{-32}$ g.

Utilizing some data obtained in observations of pulsars, FEINBERG [*123*] concluded that the only dispersion occurring in the radio to optical regions is given by $n^2 = 1 - \omega_p^2/\omega^2$. Although he was looking into the problem of the photon mass in vacuo, a task entirely different from ours, his results are directly applicable here.

Whilst explaining the FREUNDLICH [*130*] red shift, TER HAAR [*131*] concluded that the effect of electrons on the passing electromagnetic wave is stronger than the corresponding effect caused by photons.

There is still another effect observed on the astronomical scale, viz., that of quasar red shifts which are not caused by the Doppler velocities [*132*]. In order to explain this phenomenon, PECKER et al. [*133*] introduced the concept of an inelastic photon-photon interaction. They claim that a redshift results from the slowing down of emitted photons by the radiation field surrounding the source [*134*]. Putting it differently, they suggest that the interaction of a transverse background photon with a (transverse) emitted photon may be mediated by a longitudinal photon — thus assuming that photons should have non-vanishing rest masses. This explanation was queried by WOODWARD and YOURGRAU [*135*] as not fully meeting the requirements of the "tired light" hypothesis.

Another anomalous red shift effect was observed by GOLDSTEIN [*136*], when the satellite Pioneer-6 reached its perihelion behind the Sun. MERAT et al. [*137*] have analysed this shift, which was symmetrical with respect to the center of the Sun, and concluded that the effect was due to the interaction of the photons emitted by Pioneer-6 with the radiation emanating from the Sun. These anomalous shifts $\delta\nu$, and the associated energy losses, could be interpreted, so it was suggested, as being caused by the non-elastic photon-photon scattering.

In the just discussed explanations of the non-velocity red shifts, it was assumed that the mass of a photon, although minute, was still finite. However, the effect of the interstellar medium was not taken into account. Thus, a modified explanation can be put forward. As the collisionless plasma is a good approximation [*138*] in analyzing the wave phenomena in the interstellar space, it can be asserted that a photon (originally massless) acquires a non-vanishing mass when propagating through such a medium. And only then it interacts in the way proposed above. This last reasoning is in agreement with the supposition of this work.

Turn the attention now to the energy-momentum tensor for the free massive field. Recalling that the free field is one with no charges or currents present, we see that (6.11) becomes simply

$$\Box A^\mu + \mu^2 A^\mu = 0, \tag{6.16}$$

or, written in terms of the electromagnetic field tensor, we have

$$\partial_\mu F^{\mu\nu} + \mu^2 A^\nu = 0. \tag{6.17}$$

As a consequence of the mutual dependence of A^μ and $F^{\mu\nu}$ which stems from (6.17) and (2.4), the $\mu \neq 0$ theory cannot be gauge invariant [*48*]. This needs an explanation, since gauge invariance is an important concept. The vector potential, introduced as an auxiliary mathematical field to help to simplify the analysis of physical field, is not directly measurable. However, in order that correct values of field strengths are obtained, one may always add gradient of a scalar Λ without affecting those, directly measurable, field variables. The vector potential itself, however, is not uniquely defined. The transformation

$$A^\mu \to A^\mu + \partial^\mu \Lambda \tag{6.18}$$

is then termed the gauge transformation. Hence, in order that a field quantity has a physical meaning, it must be invariant under gauge transformations (6.18). In this connection, we may recall the idea of AHARONOV and BOHM [*111, 112*] to consider the vector potentials A^μ (rather than the field tensor $F^{\mu\nu}$) as the true physical field quantities. It can be shown [*125*] that the theoretical requirement of the gauge invariance implies the vanishing photon mass. Nevertheless, COESTER [*139*] was able to construct a theory of a neutral vector meson which, whilst not being gauge invariant, remained in accord with observations.

Now, as is well known, the energy-momentum tensor can be formed from the Lagrange density. Thus, if the Hamiltonian density is defined by

$$\mathscr{H}_P = \frac{\partial \mathscr{L}_P}{\partial(\partial_t A_\alpha)}\, \partial_t A_\alpha - \mathscr{L}_P, \tag{6.19}$$

the covariant formulation of (6.19) leads (in view of the Lorentz transformation properties of $\mathscr{H}$) to the canonical energy-momentum density tensor (2.28). Substitution of the Proca Lagrangian density (6.9) into (2.28) and utilizing the corresponding field equations (6.12), yields the following expressions for components of this tensor

$$\begin{aligned} T^{00} &= \frac{1}{8\pi}\,[(E^2 + H^2) + \mu^2(V^2 + A^2)] + \frac{1}{4\pi}\,\nabla \cdot (V\boldsymbol{E}) \\ T^{0i} &= \frac{1}{4\pi}\,[(\boldsymbol{E}\times\boldsymbol{H})_i + \mu^2 A_i V + \nabla \cdot (A_i \boldsymbol{E})] \\ T^{i0} &= \frac{1}{4\pi}\,[(\boldsymbol{E}\times\boldsymbol{H})_i + \mu^2 A_i V + (\nabla + V\boldsymbol{H})_i - \partial_t(VE_i)]. \end{aligned} \tag{6.20}$$

Generally speaking, the energy density corresponding to (6.9) is not positive definite [*48*].

Components T^{00} and T^{0i} in (6.20), corresponding to the energy and momentum densities, respectively, are clearly seen to differ from the accepted ones. Nevertheless (cf. Chapter II), by a three dimensional space intergation, the following expressions for the energy and momentum, associated with the Proca field, are obtained

$$\begin{aligned} \int T^{00} d^3x &= \frac{1}{8\pi} \int [E^2 + H^2 + \mu^2(V^2 + A^2)]\, d^3x \\ \int T^{0i} d^3x &= \frac{1}{4\pi} \int [(\boldsymbol{E}\times\boldsymbol{H})_i + \mu^2 A_i V]\, d^3x. \end{aligned} \tag{6.21}$$

The shortcomings of the canonical tensor were discussed in Chapter II. In an attempt to meet them, BELINFANTE [*46*] proposed a symmetrical form for the tensor density. Thus, the total angular momentum density tensor $J^{\lambda\mu\nu}$ can be written as

$$J^{\lambda\mu\nu} = M^{\lambda\mu\nu} + N^{\lambda\mu\nu}, \tag{6.22}$$

a sum of an orbital $M^{\lambda\mu\nu}$ and spin $N^{\lambda\mu\nu}$ angular momentum density tensors. However, this last subdivision may be shown to be unique only when the rest frame of the particle can be defined. Thus, a meson, having a non-vanishing rest mass, satisfies this condition. Therefore, it lends itself for the analysis of the energy-momentum tensor of the electromagnetic field in a plasma.

Our primary concern in defining the symmetrical energy-momentum tensor $\Theta^{\mu\nu}$ is to modify $T^{\mu\nu}$ in such a way that the total distribution of energy and momentum densities in the field, described by $\Theta^{\mu\nu}$, be given by

$$\Theta^{\mu\nu} = T^{\mu\nu} + T_s^{\mu\nu}. \tag{6.23}$$

A divergenceless tensor $T_s^{\mu\nu}$ may be then interpreted as the spin energy-momentum tensor, and must be chosen in such a way that conservation laws involving $T^{\mu\nu}$ still hold. It is then found [*48*] that

$$T_s^{\mu\nu} = \frac{1}{4\pi}\,\partial_\lambda F^{\mu\lambda} A^\nu. \tag{6.24}$$

It is to be noted that $T_s^{\mu\nu}$ gives no contribution to the total energy and momentum [*46*], since

$$\int \Theta^{\lambda_0} d^3x = \int T^{\lambda_0} d^3x. \tag{6.25}$$

The total angular momentum density tensor

$$J^{\lambda\mu\nu} = x^\lambda \Theta^{\mu\nu} - x^\mu \Theta^{\lambda\nu} \tag{6.26}$$

is now conserved, as its divergence, $\partial_\nu J^{\lambda\mu\nu}$, vanishes. Thus, the contribution of $T_s^{\mu\nu}$ ensures the conservation of $J^{\lambda\mu\nu}$.

To elucidate the meaning of (6.23), we express it in a component form, making use of the Proca equations (6.12), as

$$\begin{aligned} \Theta^{00} &= \frac{1}{8\pi}\,[E^2 + H^2 + \mu^2(V^2 + A^2)] \\ \Theta^{i0} &= \Theta^{0i} = \frac{1}{4\pi}\,[(\boldsymbol{E}\times\boldsymbol{H})_i + \mu^2 A_i V]. \end{aligned} \tag{6.27}$$

Similar expressions for the massive vector field were obtained by SCHRÖDINGER [*127*]. At least, for a relatively loose coupling between the tensor and the potential fields, there will be no mixed terms in (6.27). As μ is assumed to be small, the above expressions very closely resemble the ordinary Maxwell field.

There is one other point worth mentioning. Using the field equations (6.12), the linear momentum derived from the canonical tensor $T^{\mu\nu}$ becomes

$$g = \frac{1}{4\pi}\int V(\boldsymbol{E}\cdot\boldsymbol{A})\,d^3x. \tag{6.28}$$

The corresponding expression, obtained from the symmetric tensor $\Theta^{\mu\nu}$, takes the form

$$g = \frac{1}{4\pi}\int \{\boldsymbol{A}(\nabla\cdot\boldsymbol{E}) + (\boldsymbol{E}\times\boldsymbol{H})\}\,d^3x. \tag{6.29}$$

The integrated total angular momentum can be written, therefore, as

$$\boldsymbol{J} = \frac{1}{4\pi}\int \boldsymbol{x}\times\{\boldsymbol{A}(\nabla\cdot\boldsymbol{E}) + (\boldsymbol{E}\times\boldsymbol{H})\}\,d^3x. \tag{6.30}$$

Using standard vector relations [50], the integrand of (6.30) assumes the form

$$\boldsymbol{x} \times \{\nabla(\boldsymbol{E} \cdot \boldsymbol{A}) - (\boldsymbol{E} \cdot \nabla) \boldsymbol{A} + \boldsymbol{A}(\nabla \cdot \boldsymbol{E}). \tag{6.31}$$

Now, in view of (6.28), the above is interpreted as the moment of the canonical momentum density and the spin momentum density. Consequently, substitution of (6.31) into (6.30) followed by an integration, results in

$$\boldsymbol{J} = \frac{1}{4\pi} \int \boldsymbol{E} \cdot (\boldsymbol{x} \times \nabla) \boldsymbol{A} d^3x + \frac{1}{4\pi} \int (\boldsymbol{E} \times \boldsymbol{A}) \, d^3x, \tag{6.32}$$

in accord with the relevant expression of BOPP [49].

This point, essentially, concludes the derivation of the field energy-momentum tensor for the neutral vector meson in vacuum. Thus, it is found that the total tensor is symmetric and consists of an "orbital" part, which determines the energy and momentum, and a „spin" part, which does not contribute to either, but is important in calculations of the total angular momentum.

A remark about the energy-momentum tensor (6.27) is in order here. The time component of the vector field (V, $\boldsymbol{A}$) vanishes in the case of a transverse field, as was noted in comments following (6.12). Therefore, its contribution to Θ^{00} in (6.27) is nil. Moreover, the space-time component of $\Theta^{\mu\nu}$ is now $[\boldsymbol{E} \times \boldsymbol{H}]/4\pi$ and the wave is, therefore, nearly, pure transverse wave.

Hence, in accordance with the thesis of this work, the same conclusion must also apply to an electromagnetic field in an isotropic, collisionless plasma. Put otherwise, in order that a tensor have a physical meaning, it must be symmetric.

Next, it is well accepted [18, 23, 78] that the Minkowski tensor $S_M{}^{\mu\nu}$ readily adjusts itself to the canonical formalism. Recall, further, that that tensor (similarly to the canonical tensor) is not symmetric. Thus, in analogy to the above discussion, the tensor $S_{sp}^{\mu\nu} = S_A{}^{\mu\nu} - S_M{}^{\mu\nu}$ ought to be interpreted as the spin energy-momentum tensor. Expressed covariantly, it becomes

$$S_{sp}^{\mu\nu} = (\varepsilon\mu - 1) \, g^{\mu\alpha} \Omega_\alpha v^\nu \tag{6.33}$$

and, in the three dimensional form in the rest frame

$$\begin{aligned} S_{sp}^{00} &= 0, \qquad S_{sp}^{0i} = 0, \\ S_{sp}^{i0} &= \frac{\varepsilon\mu - 1}{4\pi c} \boldsymbol{E} \times \boldsymbol{H}. \end{aligned} \tag{6.34}$$

It follows, from (6.34) that the only non-vanishing component of $S_{sp}^{\mu\nu}$ is the space-time one. Let us inquire, then, about its effect on the total tensor which, in view of the definition of $S_{sp}^{\mu\nu}$, is identical with the Abraham form $S_A{}^{\mu\nu}$. Recalling the classical meson theory, presented above, we see that all high frequency phenomena can be approximately described by the Maxwell theory. Put otherwise, the characteristic differences between two theories become pronounced only in the low frequency limit.

A somewhat similar conclusion can be reached when investigating the propagation of the electromagnetic radiation in a medium. In the process, this medium becomes, generally speaking, magnetized. On a microscopic level, this effect corresponds to the spin of the field quanta. Furthermore, it is known [8] that the magnetic effects are much smaller than dielectric effects and that they belong, essentially, to a low frequency domain. Let us apply now this consideration to the force corresponding to (6.34). Thus, in the rest frame, the corresponding expression takes the form

$$f_{sp} = \frac{\varepsilon\mu - 1}{4\pi c} \partial_t(\boldsymbol{E} \times \boldsymbol{H}). \tag{6.35}$$

Now, recall that, in the low frequency domain, the value of the dielectric permittivity is considerably different from that of the vacuum. On the other hand, it is well accepted that the value of the permittivity, at high frequencies, asymptotically approaches unity, as a glance at (4.25) will reveal. Therefore, this behaviour also affects the magnitude of f_{sp}. Thus, it may be deduced that the magnitude of f_{sp} will almost vanish at high frequencies, whilst it will increases for lower frequencies.
Although this work concentrated mostly on the case of a collisionless, isotropic plasma, nevertheless, the above interpretation is in excellent agreement with the experimental results obtained by using liquid and solid dielectrics. Thus, the conclusion reached here predicts a vanishing spin energy-momentum tensor (6.33) in the region of high frequencies. Therefore, in view of the above discussion, the field energy-momentum tensor is quite correctly given by the Minkowski theory, when the frequency is high. This conclusion is well supported by experiments [*25, 26, 27*]. At the other end of the frequency spectrum, (6.33) does not vanish and, thus, the field energy- momentum tensor is then a sum of $S_M{}^{\mu\nu}$ and $S_{sp}^{\mu\nu}$, directly leading to the form put forward by ABRAHAM. The experimental evidence supplied by [*28, 29, 30*] well supports this conclusion.
This concludes our discussion of the Proca theory of neutral vector mesons and of its applications to the problem of the electromagnetic energy-momentum tensor in a medium. In fact, we have, essentially, completed the task attempted in this work. It only remains for us to summarize and synthesize our findings, placing them in proper perspective. This will be done in the final chapter, where also some extensions for future work will be suggested.

VII. Conclusion

In the preceding chapters, we dealt with a few aspects of the interaction phenomena between the electromagnetic wave and a medium. Our main effort was devoted to deducing the correct expression for the field energy-momentum tensor in a medium. As discussed in Chapter III, many different suggestions had been previously put forward and we showed that the emerging picture reflected a rather confused situation with no consensus of opinions as to which form of that tensor was correct. This was shown to be partly due to some very different interpretations being given to certain physical quantities.
In view of the complexity of the problem, it was considered necessary to make a few approximations, in order to elucidate the physics of the question at hand. Thus, the main effort was directed to the case of an isotropic, collisionless plasma, i.e., a collisionless plasma in the absence of an external magnetic field. Although this resulted in a rather artificial problem, it helped to clarify the distinction between the suggested tensors. Then, as a further step towards generalization, collisions could be included, as the above method is readily adaptable.
According to the present writer's knowledge, all relevant works, hitherto, deal with the interaction of the electromagnetic field with a medium whose refractive index is greater than unity. It was, therefore, considered essential to concentrate instead on interactions with media whose refractive index is smaller than unity. Although the most obvious example of such a medium is a plasma, we know that similar considerations also apply to a gas, liquid, or a solid, provided that the frequency of the incident radiation is very much larger than the characteristic frequency of a medium.

It proved illustrative to consider the radiation from both points of view — those of corpuscules and waves. The differentiation between these two aspects turned out to be more profound when radiation was inside a medium. Within the boundaries of geometrical optics approximations, it was found (in Chapter V) that the momentum of a photon in a plasma is given by the Minkowski expression. In the process, the method of Einstein, to investigate the interaction of radiation with matter, was expanded and used, in turn, to support the Minkowski result. It should be emphasized, however, that this conclusion is valid only in the region of high frequencies. There is some experimental evidence, viz., that of JONES and RICHARDS [25], JONES and LESLIE [26], and ASHKIN and DZIEDZIC [27] which tends to support this finding.

To investigate the behaviour of the low frequency electromagnetic field in a medium, the similarity of the relativistic form of the energy-momentum relation for a massive particle to the dispersion equation for the wave motion was invoked. Although the preliminary steps were introduced earlier [40, 43], the resulting neutral vector meson theory had not been previously applied to the problem of the form of the energy-momentum tensor. It should be emphasized that, although the theory itself is not new, the underlying philosophy is certainly entirely novel. Furthermore, it should be stressed that, throughout this work, the mass of the photon in vacuo is assumed to be zero. However, when the radiation field propagates in a medium, it may be represented by photons having non-zero rest mass. The coupling between the radiation field and a medium is responsible for the existence of the massive photons-like hehaviour.

Therefore, the meson theory, together with the symmetrization procedure of BELINFANTE [46], was utilized to determine the form of the energy-momentum tensor. The work of BOPP [49] was instrumental in drawing an analogy between the tensors $T^{\mu\nu}$ and $\Theta^{\mu\nu}$ of classical electrodynamics, on one hand, and those suggested by Minkowski and Abraham, on the other. It was then found that the Abraham form of the tensor is, generally, the correct one. The Minkowski tensor was definitely ruled out in the low frequency limit. This agrees with results of the work of WALKER et al. [28], and WALKER and WALKER [29]. Further, the results obtained by HAKIM and HIGHAM [30], in the static limit, definitely supported the Abraham conclusion. (It is instructive to notice that in experiments [28, 29], the used frequencies were in the range 5—100 Hz and, thus, obviously, did not satisfy the WKB-condition). Although SHOCKLEY and JAMES [54] did not actually conclude that the value of the momentum was that given by the Abraham expression, they, nevertheless, definitely excluded the Minkowski result, while using the frequency of 5 kHz.

The formal equivalence of the meson theory in vacuo with the Maxwell theory in a medium was derived while utilizing the concept of a collisionless plasma. The first step towards generalization would be to apply the above method to the case of a collisional plasma. No real difficulty is envisaged and, it appears that, the extension would be straightforward. Further, although there is qualitative agreement with the experiment, it would be informative to extend the above method to apply to any state of matter. Moreover, the interaction of a high density field, e.g., that of laser, with a medium would certainly warrant further investigation.

To summarize, the correct form for the field-energy momentum tensor has been given by Abraham. The Minkowski form for that tensor, on the other hand, appears to represent a fair approximation in the region of high frequencies, where the spin contributions may be neglected.

Acknowledgements

The author would like to express his gratitude to —

Professor H. Hora, who introduced him to the subject and whose constant enthusiasm made this investigation very rewarding;

J. R. Shepanski, who worked through the manuscript and whose comments proved to be of great value;

D. T. Holmes, who conscientiously read the following pages and contributed towards an early completion of this work.

Bibliography

[1] E. Schrödinger, Proc. Royal Soc. **A 232**, 435 (1955).
[2] W. Pauli, Theory of Relativity, Pergamon Press, 1958.
[3] L. de Broglie, Phil. Mag. **47**, 446 (1924).
[4] O. Costa de Beauregard, Précis of Special Relativity, Academic Press, New York, 1966.
[5] L. D. Landau, E. M. Lifshitz, The Classical Theory of Fields, Pergamon Press, 1971.
[6] M. von Laue, Z. Phys. **128**, 387 (1950).
[7] C. Moller, The Theory of Relativity, Oxford, 1952.
[8] L. D. Landau, E. M. Lifshitz, Electrodynamics of Continuous Media, Pergamon, 1960.
[9] N. L. Balazs, Phys. Rev. **91**, 408 (1953).
[10] D. V. Skobel'tsyn, Uspekhi Fiz. Nauk **110**, 253 (1973).
[11] M. G. Burt, R. Peierls, Proc. Royal Soc. **A 333**, 149 (1973).
[12] J. P. Gordon, Phys. Rev. **A 8**, 14 (1973).
[13] R. Peierls, Proc. Royal Soc. **A 347**, 475 (1976).
[14] R. V. Jones, Proc. Royal Soc. **A 360**, 365 (1978).
[15] W. Shockley, Proc. N. A. S. **60**, 807 (1968).
[16] H. A. Haus, Physica **43**, 77 (1969).
[17] F. N. H. Robinson, Phys. Rep. **16**, 313 (1975).
[18] I. Brevik, Mat. Fys. Medd. Dan. Vid. Selsk. **37**, No. 11 and 13 (1970).
[19] V. L. Ginzburg, Soviet Physics-Uspekhi **16**, 434 (1973).
[20] Z. Mikura, Phys. Rev. **A 13**, 2265 (1976).
[21] A. Einstein, J. Laub, Ann. Phys. (Leipzig) **26**, 532, 541 (1908); **27**, 232 (1908).
[22] F. Beck, Z. Phys. **134**, 136 (1953).
[23] G. Marx, G. Györgyi, Ann. Phys. (Leipzig) **16**, 241 (1955).
[24] S. R. de Groot, L. G. Suttorp, Physica **37**, 284, 297, (1967); **39**, 28, 41, 61, 77, 84 (1968).
[25] R. V. Jones, J. C. S. Richards, Proc. Royal Soc. **A 221**, 480 (1954).
[26] R. V. Jones, B. Leslie, Proc. Royal Soc. **A 360**, 347 (1978).
[27] A. Ashkin, J. M. Dziedzic, Phys. Rev. Letters **30**, 139 (1973).
[28] G. B. Walker, D. G. Lahoz, G. Walker, Canad. J. Phys. **53**, 2577 (1975).
[29] G. B. Walker, G. Walker, Canad. J. Phys. **55**, 2121 (1977).
[30] S. S. Hakim, J. B. Higham, Proc. Phys. Soc. **80**, 190 (1962).
[31] C. G. Darwin, Proc. Royal Soc. **A 146**, 17 (1934).
[32] B. B. Kadomtsev, Soviet Physics JETP **6**, 117 (1958).
[33] V. L. Ginzburg, The Propagation of Electromagnetic Waves in Plasma, Pergamon, 1964.
[34] A. Einstein, Phys. Z. **18**, 121 (1917).
[35] D. V. Skobel'tsyn, Voprosy Matematicheskoi Fiziki, Nauka, Leningrad, 1976.
[36] G. Bekefi, Radiation Processes in Plasma, Wiley, 1966.
[37] J. M. Dawson, Adv. Plasmas Phys. **1**, 1 (1968).
[38] K. D. Cole, Austral. J. Phys. **30**, 671 (1977).
[39] H. Hora, Phys. Fluids **12**, 182 (1969).
[40] K. D. Cole, Austral. J. Phys. **24**, 871 (1971); **26**, 359 (1973).
[41] K. D. Cole, Phys. Letters **39 A**, 278 (1972); **41 A**, 325 (1972).
[42] A. R. Lee, Phys. Letters **36 A**, 283 (1971); **41 A**, 323 (1972).

[43] R. KLIMA, V. A. PETRZILKA, Czech. J. Phys. **B 23**, 1196 (1973).
[44] H. C. CORBEN, J. SCHWINGER, Phys. Rev. **58**, 953 (1940).
[45] F. ROHRLICH, Classical Charged Particles, Addison-Wesley, 1965.
[46] F. J. BELINFANTE, Physica **6**, 887 (1939).
[47] F. J. BELINFANTE, Physica **7**, 449 (1940).
[48] W. PAULI, Rev. Mod. Phys. **13**, 203 (1941).
[49] F. BOPP, Ann. Phys. (Leipzig) **11**, 35 (1963).
[50] J. D. JACKSON, Classical Electrodynamics, Wiley, 1975.
[51] F. N. H. ROBINSON, Macroscopic Electromagnetism, Pergamon, 1973.
[52] R. GRAMMEL, Ann. Phys. (Leipzig) **41**, 570 (1913).
[53] H. G. SCHÖPF, Ann. Phys. (Leipzig) **9**, 301 (1962).
[54] W. SHOCKLEY, R. P. JAMES, Phys. Rev. Letters **18**, 876 (1967).
[55] R. P. JAMES, Proc. N.A.S. **61**, 1149 (1968).
[56] O. COSTA DE BEAUREGARD, Phys. Letters **21**, 410 (1966).
[57] CH. GOILLOT, Phys. Letters **21**, 408 (1966).
[58] O. COSTA DE BEAUREGARD, Nuovo Cimento **48 B**, 293 (1967).
[59] O. COSTA DE BEAUREGARD, Phys. Letters **24 A**, 177 (1967).
[60] O. COSTA DE BEAUREGARD, Found. Physics **2**, 111 (1972).
[61] A. PAPAPETROU, Phil. Mag. **40**, 937 (1949).
[62] G. GYÖRGYI, Amer. J. Phys. **28**, 85 (1960).
[63] J. AGUDIN, Phys. Letters **24 A**, 761 (1967).
[64] D. V. SKOBEL'TSYN, Uspekhi Fiz. Nauk **122**, 295 (1977).
[65] R. L. DEWAR, Phys. Fluids **13**, 2710 (1970).
[66] W. L. JONES, Rev. Geophys. Space Phys. **9**, 917 (1971).
[67] M. VON LAUE, Das Relativitätsprinzip, Braunschweig, 1911.
[68] A. SCHEYE, Ann. Phys. (Leipzig) **30**, 805 (1909).
[69] C. L. TANG, J. MEIXNER, Phys. Fluids **4**, 148 (1961).
[70] G. H. LIVENS, Phil. Mag. **32**, 162 (1916).
[71] R. PEIERLS, Proc. Royal Soc. **A 355**, 141 (1977).
[72] H. K. WONG, K. YOUNG, Amer. J. Phys. **45**, 195 (1977).
[73] G. A. ASKAR'YAN, JETP Letters **9**, 241 (1969).
[74] K. K. THORNBER, Phys. Letters **43 A**, 501 (1973).
[75] J. A. ARNAUD, Optics Comm. **7**, 313 (1973).
[76] W. B. JOYCE, Amer. J. Phys. **43**, 245 (1975).
[77] W. B. JOYCE, Phys. Rev. **D 9**, 3234 (1974).
[78] R. L. DEWAR, Austral. J. Phys. **30**, 533 (1977).
[79] E. L. HILL, Rev. Mod. Phys. **23**, 253 (1951).
[80] E. J. POST, Phys. Rev. **118**, 1113 (1960).
[81] C. ECKART, Phys. Rev. **54**, 920 (1938).
[82] R. KLIMA, V. A. PETRZILKA, Czech. J. Phys. **B 22**, 896 (1972).
[83] R. KLIMA, V. A. PETRZILKA, Ann. Phys. (N.Y.) **92**, 395 (1975).
[84] P. A. STURROCK, Ann. Phys. (N.Y.) **4**, 306 (1958).
[85] F. HEHL, Ann. Phys. (Leipzig) **25**, 215 (1970).
[86] F. BOPP, Ann. Phys. (Leipzig) **25**, 221 (1970).
[87] J. H. POYNTING, Phil. Mag. **9**, 393 (1905).
[88] G. BARLOW, Proc. Royal Soc. **A 87**, 1 (1912).
[89] R. A. BETH, Phys. Rev. **50**, 115 (1936).
[90] N. CARRARA, Nature **164**, 882 (1949).
[91] E. G. RAWSON, A. D. MAY, Appl. Phys. Letters **8**, 93 (1966).
[92] A. D. MAY, E. G. RAWSON, E. H. HARA, J. Appl. Phys. **38**, 5290 (1967).
[93] A. ASHKIN, Phys. Rev. Letters **24**, 156 (1970).
[94] H. M. LAI, K. YOUNG, Phys. Rev. **A 14**, 2329 (1976).
[95] A. V. KATS, V. M. KONTOROVICH, JETP Letters **9**, 112 (1969).
[96] J. F. DENISSE, J. L. DELCROIX, Plasma Waves, Interscience, New York, 1963.
[97] L. SPITZER, Physics of Fully Ionized Gases, Interscience, New York, 1956.
[98] J. H. PIDDINGTON, Phil. Mag. **47**, 1037 (1955).
[99] D. R. HARTREE, Proc. Cambridge Phil. Soc. **25**, 47 (1929); **27**, 143 (1932).

[*100*] K. C. WESTFOLD, Phil. Mag. **41**, 509 (1950).
[*101*] L. BRILLOUIN, Wave Propagation and Group Velocity, Academic Press, New York, 1960.
[*102*] W. HEITLER, The Quantum Theory of Radiation, Oxford, 1954.
[*103*] H. HORA, Phys. Fluids **17**, 1042 (1974).
[*104*] D. PARK, Ann. Phys. (Leipzig) **27**, 319 (1971).
[*105*] R. M. LEWIS, Arch. Ration. Mech. Analysis **20**, 191 (1965).
[*106*] L. DE BROGLIE, Non-linear Wave Mechanics, Elsevier, Amsterdam, 1960.
[*107*] V. A. FOCK, Electromagnetic Diffraction and Propagation Problems, Pergamon Press, 1965.
[*108*] J. A. ARNAUD, J. Opt. Soc. Amer. **65**, 174 (1975).
[*109*] H. HORA, Optik **17**, 409 (1960).
[*110*] J. L. CARTER, H. HORA, J. Opt. Soc. Amer. **61**, 1640 (1971).
[*111*] Y. AHARONOV, D. BOHM, Phys. Rev. **115**, 485 (1959).
[*112*] Y. AHARONOV, D. BOHM, Phys. Rev. **123**, 1511 (1961).
[*113*] R. G. CHAMBERS, Phys. Rev. Letters **5**, 3 (1960).
[*114*] T. H. BOYER, Phys. Rev. **D 8**, 1667, 1679, (1973).
[*115*] J. DAWSON, C. OBERMAN, Phys. Fluids **5**, 517 (1962).
[*116*] L. OSTER, Astrophys. J. **138**, 761 (1963); **144**, 836 (1966).
[*117*] K. M. CASE, S. C. CHIU, Phys. Rev. **A 1**, 1170 (1970).
[*118*] W. M. CRONYN, Astrophys. J. **144**, 834 (1966).
[*119*] D. F. MARTYN, Proc. R. Soc. **A 193**, 44 (1948).
[*120*] C. L. INMAN, Astrophys. J. **142**, 201 (1965).
[*121*] R. M. MORE, J. Phys. **A 9**, 1979 (1976).
[*122*] V. A. PETRZILKA, Czech. J. Phys. **B 26**, 151 (1876).
[*123*] G. FEINBERG, Science **166**, 879 (1969).
[*124*] A. S. GOLDHABER, M. M. NIETO, Rev. Mod. Phys. **43**, 377 (1971).
[*125*] J. M. JAUCH, F. ROHRLICH, The Theory of Photons and Electrons, Addison-Wesley, 1955.
[*126*] M. A. GINTSBURG, Soviet Astron. AJ **7**, 536 (1964).
[*127*] L. BASS, E. SCHRÖDINGER, Proc. Royal Soc. **A 232**, 1 (1955).
[*128*] I. YU. KOBZAREV, L. B. OKUN, Soviet Physics-Uspekhi **11**, 338 (1968).
[*129*] M. E. GERTSENSHTEIN, L. G. SOLOVEI, JETP Letters **9**, 79 (1969).
[*130*] E. FINLAY-FREUNDLICH, Phil. Mag. **45**, 303 (1954).
[*131*] D. TER HAAR, Phil. Mag. **45**, 320 (1954).
[*132*] H. ARP, Science **174**, 1189 (1971).
[*133*] J. C. PECKER, A. P. ROBERTS, J. P. VIGIER, Nature 237, 227 (1972).
[*134*] J. C. PECKER, W. TAIT, J. P. VIGIER, Nature **241**, 338 (1973).
[*135*] J. F. WOODWARD, W. YOURGRAU, Nature **241**, 338 (1973).
[*136*] R. M. GOLDSTEIN, Science **166**, 598 (1969).
[*137*] P. MERAT, J. C. PECKER, J. P. VIGIER, Astron. Astrophys. **30**, 167 (1974).
[*138*] I. LERCHE, Adv. Plasma Phys. **2**, 47 (1969).
[*139*] F. COESTER, Phys. Rev. **83**, 798 (1951).

INHALT

Seite

M. M. NOVAK: Interactions of Photons with Electrons in Dielectric Media 285

Die „FORTSCHRITTE DER PHYSIK“ dienen der Publikation zusammenfassender Berichte über aktuelle Teilgebiete der Physik. Neben Originalberichten in deutscher oder englischer Sprache werden deutsche Übersetzungen wichtiger fremdsprachiger Berichte abgedruckt.

Aus dem Inhalt der folgenden Hefte:

FUSCO-GIRARD, M., F. MANCINI, M. MARINARO: Electrodynamics of Superconductors as a Consequence of Local Gauge Invariance

BARBASHOV, B. M., V. V. NESTERENKO: Differential Geometry and Nonlinear Field Models

KAZAKOV, D. I., D. V. SHIRKOV: Asymptotic Series of Quantum Field Theory and Their Summation

DARBAIDZE, YA. Z., S. M. ESAKIA, V. R. GARSENASHVILI, Z. R. MENTESHASHVILI: Problems of Deep Inelastic Lepton-Nucleus Interaction

GRASSBERGER, P.: The Gribov Process: Soft Parton Interactions in So-called Reggeon Field Theory

Manuskripte werden nach Vereinbarung mit dem Autor von der Schriftleitung entgegengenommen.

MIX
Papier | Fördert gute Waldnutzung
FSC® C083411

Zeitfracht Medien GmbH
Ferdinand-Jühlke-Straße 7
99095 Erfurt, Deutschland
produktsicherheit@kolibri360.de